Maths Problem Solving

Year 1

Catherine Yemm

Brilliant Publications

Other books in the same series:

Maths Problem Solving – Year 2	ISBN 10…1-903853-74-3
	ISBN 13… 978-1-903853-75-7
Maths Problem Solving – Year 3	ISBN 10… 1-903853-76-1
	ISBN 13… 978-1-903853-76-4
Maths Problem Solving – Year 4	ISBN 10… 1-903853-77-X
	ISBN 13… 978-1-903853-77-1
Maths Problem Solving – Year 5	ISBN 10… 1-903853-78-8
	ISBN 13… 978-1-903853-78-8
Maths Problem Solving – Year 6	ISBN 10… 1-903853-79-6
	ISBN 13… 978-1-903853-79-5

Published by Brilliant Publications
1 Church View, Sparrow Hall Farm, Edlesborough,
Dunstable, Bedfordshire LU6 2ES

Sales and stock enquiries	Tel:	0845 1309200 / 01202 712910
	Fax:	0845 1309300
Sales and payments	E-mail:	brilliant@bebc.co.uk
	website:	www.brilliantpublications.co.uk
General enquiries:	Tel:	01525 229720

The name Brilliant Publications and its logo are registered trade marks.

Written by Catherine Yemm
Cover and illustrations by Frank Endersby

ISBN 10… 1-903853-74-5
ISBN 13… 978-1-903853-74-0
© Catherine Yemm 2005
First published in 2005
Printed in the UK by Lightning Source
10 9 8 7 6 5 4 3 2

The right of Catherine Yemm to be identified as the author of this work has been asserted by her in accordance with the Copyright, Design and Patents Act 1988.

Pages 10–106 may be photocopied by individual teachers for class use, without prior permission from the publisher. The materials may not be reproduced in any other form or for any other purpose without the prior permission of the publisher.

Contents

Introduction 4–9
Photocopiable answer sheet 10

Making decisions			**11–34**
Lesson 1	Autumn term, 1st half	Units 2–4	11–14
Lesson 2	Autumn term, 2nd half	Units 9–11	15–18
Lesson 3	Spring term, 1st half	Units 2–4	19–22
Lesson 4	Spring term, 2nd half	Units 9–10	23–26
Lesson 5	Summer term, 1st half	Units 2–4	27–30
Lesson 6	Summer term, 2nd half	Units 9–11	31–34

Reasoning about numbers or shapes			**35–58**
Lesson 1 (Shape 1)	Autumn term, 1st half	Units 5–6	35–38
Lesson 2 (Number 1)	Autumn term, 2nd half	Unit 8	39–42
Lesson 3 (Shape 2)	Spring term, 1st half	Units 5–6	43–46
Lesson 4 (Number 2)	Spring term, 2nd half	Unit 8	47–50
Lesson 5 (Shape 3)	Summer term, 1st half	Units 5–6	51–54
Lesson 6 (Number 3)	Summer term, 2nd half	Unit 8	55–58

Problems involving 'real life', money or measures			**59–82**
Lesson 1	Autumn term, 1st half	Units 2–4, 5–6	59–62
Lesson 2	Autumn term, 2nd half	Units 9–11, 12–13	63–66
Lesson 3	Spring term, 1st half	Units 2–4, 5–6	67–70
Lesson 4	Spring term, 2nd half	Units 9–10, 11–12	71–74
Lesson 5	Summer term, 1st half	Units 2–4, 5–6	75–78
Lesson 6	Summer term, 2nd half	Units 9–11, 12–13	79–82

Organizing and using data			**83–106**
Lessons 1–2	Autumn term, 2nd half	Units 12–13	83–90
Lessons 3–4	Spring term, 2nd half	Units 11–12	91–98
Lessons 5–6	Summer term, 2nd half	Units 12–13	99–106

Answers 106–109

Introduction

Maths Problem Solving Year 1 is the first book in a series of six resource books for teachers to use during the Numeracy lesson. They specifically cover the objectives from the Numeracy framework that are collated under the heading 'Solving problems'. Each book is specific for a particular year group and contains clear photocopiable resources which can be photocopied onto acetate to be viewed by the whole class or photocopied onto paper to be used by individuals.

Problem solving plays a very important part in the Numeracy curriculum and one of the reasons Numeracy is such an important subject is because the skills the children learn enable them to solve problems in other aspects of their lives. It is not enough to be able to count, recognize numbers and calculate; children need to be able to use problem solving skills alongside mathematical knowledge to help them succeed in a variety of 'real life' situations. Many of the problem solving skills and strategies that are needed do not come naturally so they have to be taught.

Problem solving is not an area which should be taught exclusively on its own but one which should be taught alongside other mathematical areas such as numbers, and shape, space and measures. Children will benefit from being given opportunities to solve problems in other areas of the curriculum and away from the classroom as well as in their Numeracy lessons.

When teaching children how to solve problems, the Numeracy strategy refers to a number of points that need to be considered:

- The length of the problems should be varied depending on the age group. Children will benefit from being given short, medium-length and more extended problems.
- The problems on one page or in one lesson should be mixed so that the children do not just assume they are all 'multiplication' problems, for example, and simply multiply the numbers they see to find each answer.
- The problems need to be varied in their complexity: there should be some one-step and some two-step problems, and the vocabulary used in each problem should differ.
- Depending on the age of the children the problems can be given orally or in writing. When given written problems to solve, some children may need help to read the words, although this does not necessarily mean that they will need help to find the answer to the question.
- The context of the problem should be meaningful and relevant to the children. It should attempt to motivate them into finding the answer and be significant to the time. For example, euros should be included as well as pounds.

This resource book is organized into four chapters: 'Making decisions', 'Reasoning about numbers or shapes', 'Problems involving "real life", money or measures' and 'Organizing and using data'. Each chapter contains six lessons, one to be used each half term.

Making decisions

The objective outlined under the 'Making decisions' heading of the Numeracy Strategy for Year 1 children is: Choose and use appropriate number operations and mental strategies to solve problems.

At this age children need to begin to realize that it will not always be obvious how to find the answer to a question. There are different operations they could use and they need to think about which operation is needed to find the answer to a particular question. The questions in this chapter require the children to add or subtract. Mixed questions are given in each activity so that the children are not left to assume that they should add or subtract each time but learn to think openly and make a decision depending on the vocabulary used and the question itself. The emphasis should be on choosing the correct operation. Each question provides the opportunity for the children to explain the operation they used to find the answer; being able to choose the appropriate operation is an important mathematical skill that needs to be developed. The questions have been written using a number of different contexts, including money and measures.

When the children are completing the questions, encourage them to think of the calculation they need to do and write it down. They should also be encouraged to record what they use to work out the answer. For example:

Ben has 3 pencils on his table and Rachel has 5 pencils on her table. How many pencils will they need to put away all together?

I will need to   3 + 5

To help me I will use   some counters

The answer is   8

Reasoning about numbers or shapes

The objectives outlined under the 'Reasoning about shapes and numbers' heading of the National Numeracy Strategy for Year 1 children are as follows:

- Solve simple mathematical problems or puzzles; recognize and predict from simple patterns and relationships. Suggest extensions by asking 'What if?' or 'What could I try next?'
- Investigate a general statement about familiar numbers or shapes by finding examples that satisfy it.
- Explain methods and reasoning orally.

The activities are a mixture of problems, puzzles and statements. Lessons 1, 3 and 5 are related to shape while lessons 2, 4 and 6 are related to number. When given a statement such as: 'If you take an odd number away from 5 you get an even number', the children should be encouraged to provide examples to prove the statement, eg **5 – 3 = 2** or **5 – 1 = 4**. Others may be more obvious questions that just need an answer. The teacher should try to spend time talking to the pupils while they are working to allow them to explain their methods and reasoning orally and to provide an opportunity to ask questions such as 'What if…?' The plenary session at the end of the lesson also provides an opportunity to do this.

Problems involving 'real life', money or measures

The objectives outlined under the 'Problems involving 'real life', money or measures' heading of the Numeracy Strategy for Year 1 children are as follows:

- Use mental strategies to solve simple problems set in 'real life', money or measurement contexts, using counting, addition, subtraction, doubling and halving, explaining methods and reasoning orally.
- Recognize coins of different values.
- Find totals and change from up to 20p.
- Work out how to pay an exact sum using smaller coins.

The activities in this chapter are typically 'word problems'. The contexts are designed to be realistic and relevant for children of Year 1 age. The questions involve the operations of adding, subtracting, doubling and halving and the questions deal with money, measurements and everyday situations. Some of the questions are one-step questions while others are simple two-step questions. The teacher should try to spend time talking to the pupils while they are working to allow them to explain their methods and reasoning orally. The plenary session at the end of the lesson also provides an opportunity to do this.

Organizing and using data

The objectives outlined under the 'Organizing and using data' heading of the Numeracy Strategy for Year 1 children are as follows:

- Solve a given problem by sorting, classifying and organizing information in simple ways, such as: using objects or pictures; in a list or simple table.
- Discuss and explain results.

As children get older they will be confronted with more and more information. Children need to be taught the skills that will enable them to order and make sense of information they collect or are given. At this age they are not ready to create their own tables in order to organize their information but can complete tables that they are given. They should have the opportunity to see data organized in different ways and to ask questions about the data.

The lesson

Mental starter

In line with the Numeracy strategy the teacher should start the lesson with a 5–10 minute mental starter. This can be practice of a specific mental skill from the list for that particular half term or ideally an objective linked to the problems the children will be solving in the main part of the lesson. For example, if the problems require the children to add and subtract then it would be useful to spend the first 10 minutes of the lesson reinforcing addition and subtraction bonds and the vocabulary involved.

The main teaching activity and pupil activity

This book aims to provide all the worksheets that the teacher will need to cover this part of the lesson successfully. The first page of each lesson, *Whole class activity*, provides three examples of the types of problems that need to be solved. They are designed to be photocopied onto acetate, alongside a blank answer sheet (see page 10), to show to the whole class using an overhead projector. The teacher will use the answer sheet to work through the examples with the class before introducing the class to the questions they can try by themselves. The teacher should demonstrate solving the problem using skills that are relevant to the abilities of the children in the class, for example using drawings, counters or number lines.

Once the children have seen a number of examples they will be ready to try some problem solving questions for themselves. Within each lesson there is a choice of three differentiated worksheets. The questions on the three worksheets are the same, but the level of mathematical complexity varies. This ensures that the questions are differentiated only according to the mathematical ability of the child. It will also ensure that when going through examples during the plenary session all the children can be involved at the same time. For example, in a question involving the addition of three numbers, the children may have had to add three different numbers, but when the teacher talks through the question the fact that to solve the problem the children need to add will be the important point being reinforced. If children are all completing totally different types of questions then when the teacher talks through a question in the plenary session some groups of children may have to sit idle as they did not have that question on their sheet. If the teacher feels that some pupils would benefit from having easier or more difficult questions then they could change the numbers on the worksheets to something more suitable.

The plenary

One of the important parts of solving problems is discussing how problems can be solved and the plenary lends itself well to this. After the children have completed the problems the plenary can be used to:

- discuss the vocabulary used in the question
- discuss how the problem can be approached
- break down a problem into smaller steps
- list the operations or calculations used to solve the problem
- discuss whether the problem can be solved in more than one way
- discuss how the answers to the problems can be checked
- divulge the answers to a number of the questions.

Support

Regardless of their mathematical ability many children of this age will find it difficult to read the questions and understand the vocabulary. Reading support should be given to the children who need it so that they are given the opportunity to practise their mathematical skills. It may also be necessary for an adult to scribe for some children.

Extension

Any children who complete their task relatively easily may need to be extended further. As well as being given the more challenging questions they could be asked to make up a question of their own, which should involve the same operations.

Resources

For some questions it will be useful to make a number of resources available to the children such as:

Counters
Number lines to 20
Multi link cubes
A selection of 2D and 3D shapes
1p, 2p, 5p, 10p and 20p coins
Analogue clocks with moveable hands.

Photocopiable answer sheet

Photocopy onto acetate and project onto wall or screen

I will need to _____

To help me I will use _____

The answer is _____

I will need to _____

To help me I will use _____

The answer is _____

I will need to _____

To help me I will use _____

The answer is _____

Making decisions

Lesson 1

Whole class activity

Ben has 3 pencils on his table and Rachel has 5 pencils on her table. How many pencils will they need to pack away altogether?

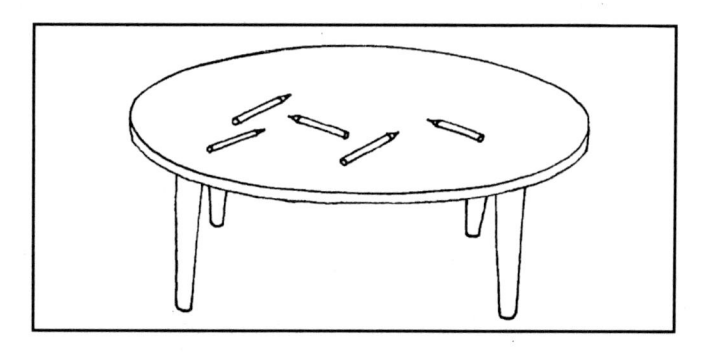

At the shop bananas cost 8p each. How much would it cost Lynn to buy 2 bananas?

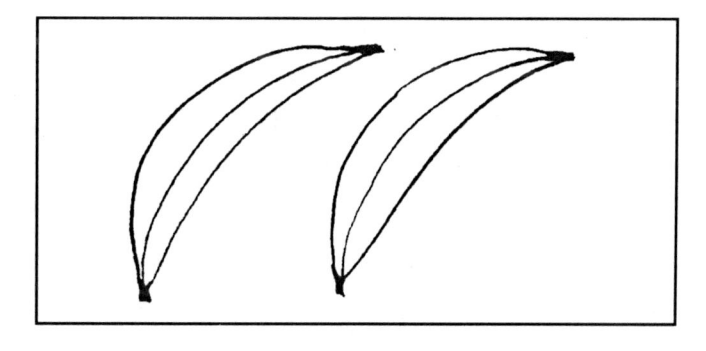

Meredith got on the bus at 3 o'clock. She got off at 5 o'clock. How long did her journey last?

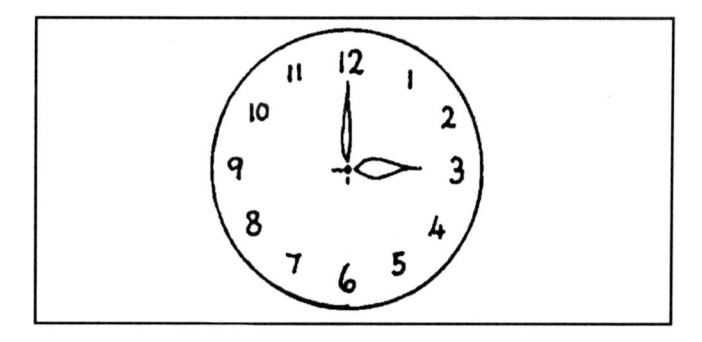

Making decisions

Lesson 1a

1. Kate has a bunch of 6 grapes in her lunchbox and eats 3 at break time. How many grapes does she have left for her lunch?

   I will need to _____

   To help me I will use _____

   The answer is _____

2. Kala's mum gives her £3 to go swimming. It costs £2 to get in the swimming pool. How much money will she have left?

   I will need to _____

   To help me I will use _____

   The answer is _____

3. George starts school at 9 o'clock. His dad starts work 1 hour earlier. What time does his dad start work?

   I will need to _____

   To help me I will use _____

   The answer is _____

4. To make her chocolate cake Haley needs 2 cups of flour. How many cups will she need if she makes 2 cakes?

   I will need to _____

   To help me I will use _____

   The answer is _____

www.brilliantpublications.co.uk — This page may be photocopied by the purchasing institution only.
Maths Problem Solving – Year 1 — © Catherine Yemm

Making decisions

Lesson 1b

1. Kate has a bunch of 10 grapes in her lunchbox and eats 6 at break time. How many grapes does she have left for her lunch?

I will need to _____

To help me I will use _____

The answer is _____

2. Kala's mum gives her £5 to go swimming. It costs £2 to get in the swimming pool. How much money will she have left?

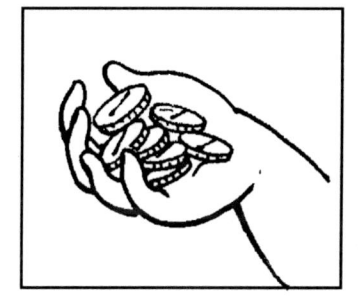

I will need to _____

To help me I will use _____

The answer is _____

3. George starts school at 9 o'clock. His dad starts work 2 hours earlier. What time does his dad start work?

I will need to _____

To help me I will use _____

The answer is _____

4. To make her chocolate cake Haley needs 4 cups of flour. How many cups will she need if she makes 2 cakes?

I will need to _____

To help me I will use _____

The answer is _____

Making decisions

1. Kate has a bunch of 15 grapes in her lunchbox and eats 8 at break time. How many grapes does she have left for her lunch?

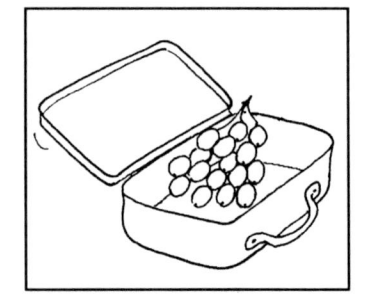

I will need to _____

To help me I will use _____

The answer is _____

2. Kala's mum gives her £10 to go swimming. It costs £4 pounds to get in the swimming pool. How much money will she have left?

I will need to _____

To help me I will use _____

The answer is _____

3. George starts school at 9 o'clock. His dad starts work 4 hours earlier. What time does his dad start work?

I will need to _____

To help me I will use _____

The answer is _____

4. To make her chocolate cake Haley needs 6 cups of flour. How many cups will she need if she makes 2 cakes?

I will need to _____

To help me I will use _____

The answer is _____

Making decisions

Whole class activity

Jacob is 7 rulers tall. Catherine is 12 rulers tall. How much taller is Catherine than Jacob?

Ashon bought a carton of juice for 6p. How much change did he have from 10p?

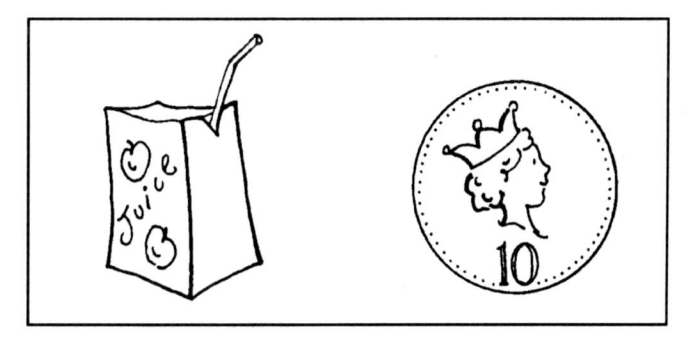

There are 12 chocolates in a box. If Julie eats half of them how many are left?

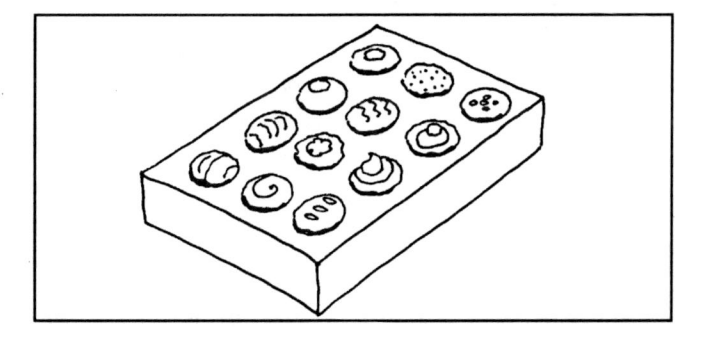

Making decisions

Lesson 2a

1. Sam has a dog and Lucy has a cat. How many legs do the pets have between them?

   I will need to _____

   To help me I will use _____

   The answer is _____

2. At lunch time Rhian would like to buy an apple for 4 pence and a drink for 6 pence. How much money will she need?

   I will need to _____

   To help me I will use _____

   The answer is _____

3. Liz's football club is 1 hour long. If it starts at 3 o'clock what time does it finish?

   I will need to _____

   To help me I will use _____

   The answer is _____

4. Craig's bucket holds 5 jugs of water and Devi's holds 9. How much more water does Devi have?

   I will need to _____

   To help me I will use _____

   The answer is _____

Making decisions

Lesson 2b

1. Sam has a dog, Lucy has a cat and Jake has a mouse. How many legs do the pets have between them?

   I will need to _____

   To help me I will use _____

   The answer is _____

2. At lunch time Rhian would like to buy an apple for 8 pence and a drink for 9 pence. How much money will she need?

   I will need to _____

   To help me I will use _____

   The answer is _____

3. Liz's football club is 2 hours long. If it starts at 3 o'clock what time does it finish?

   I will need to _____

   To help me I will use _____

   The answer is _____

4. Craig's bucket holds 8 jugs of water and Devi's holds 12. How much more water does Devi have?

   I will need to _____

   To help me I will use _____

   The answer is _____

Making decisions

Lesson 2C

1. Sam has a dog, Lucy has two cats and Jake has two mice. How many legs do the pets have between them?

   I will need to _____

   To help me I will use _____

   The answer is _____

2. At lunch time Rhian would like to buy an apple for 11 pence and a drink for 13 pence. How much money will she need?

   I will need to _____

   To help me I will use _____

   The answer is _____

3. Liz's football club is 4 hours long. If it starts at 3 o'clock what time does it finish?

   I will need to _____

   To help me I will use _____

   The answer is _____

4. Craig's bucket holds 12 jugs of water and Devi's holds 16. How much more water does Devi have?

   I will need to _____

   To help me I will use _____

   The answer is _____

Making decisions

Lesson 3

Whole class activity

The swimming pool opens at 9 o'clock and closes at 12 o'clock. How long does it stay open for?

John reads 4 pages of his book then he reads another 3 pages. How much of his book has he read?

Maya buys a cake from the cake shop for 12p. What coins could she give the shopkeeper?

Making decisions

1. Mesha needs a piece of string 4 metres long to make a kite. He has a piece of string 7 metres long. How much will he have to cut off it?

   I will need to _____

   To help me I will use _____

   The answer is _____

2. There are 5 children at the lunch table but only 2 have spoons. How many children do not have a spoon to eat their pudding?

   I will need to _____

   To help me I will use _____

   The answer is _____

3. Antony spends one hour a day riding his bike. How many hours has he spent on his bike after 4 days?

   I will need to _____

   To help me I will use _____

   The answer is _____

4. Sam weighs the same as 3 bricks, Robert weighs the same as 5 bricks. How many bricks would they weigh together if Sam jumped on Robert's back?

   I will need to _____

   To help me I will use _____

   The answer is _____

Making decisions

Lesson 3b

1. Mesha needs a piece of string 7 metres long to make a kite. He has a piece of string 12 metres long. How much will he have to cut off it?

   I will need to _____

   To help me I will use _____

   The answer is _____

2. There are 8 children at the lunch table but only 3 have spoons. How many children do not have a spoon to eat their pudding?

   I will need to _____

   To help me I will use _____

   The answer is _____

3. Antony spends one hour a day riding his bike. How many hours has he spent on his bike after 6 days?

   I will need to _____

   To help me I will use _____

   The answer is _____

4. Sam weighs the same as 7 bricks, Robert weighs the same as 8 bricks. How many bricks would they weigh together if Sam jumped on Robert's back?

   I will need to _____

   To help me I will use _____

   The answer is _____

Making decisions

1. Mesha needs a piece of string 11 metres long to make a kite. He has a piece of string 17 metres long. How much will he have to cut off it?

   I will need to _____

   To help me I will use _____

   The answer is _____

2. There are 13 children at the lunch table but only 5 have spoons. How many children do not have a spoon to eat their pudding?

   I will need to _____

   To help me I will use _____

   The answer is _____

3. Antony spends one hour a day riding his bike. How many hours has he spent on his bike after 14 days?

   I will need to _____

   To help me I will use _____

   The answer is _____

4. Sam weighs the same as 12 bricks, Robert weighs the same as 11 bricks. How many bricks would they weigh together if Sam jumped on Robert's back?

   I will need to _____

   To help me I will use _____

   The answer is _____

Making decisions

Lesson 4

Whole class activity

If it costs 8p for one child to go swimming, how much will it cost Sarah and her sister to go swimming?

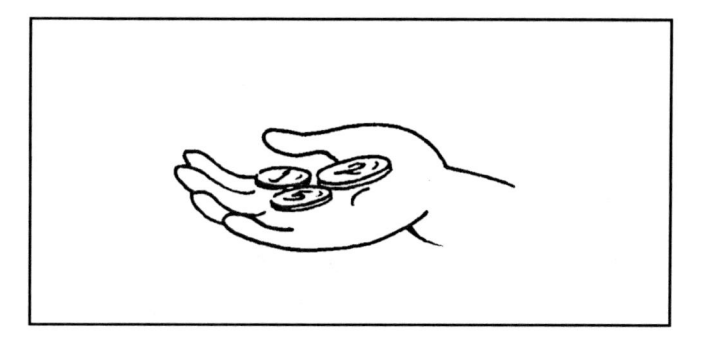

On a balance 3 marbles balance an apple and 6 marbles balance a peach. How many marbles will balance an apple and a peach?

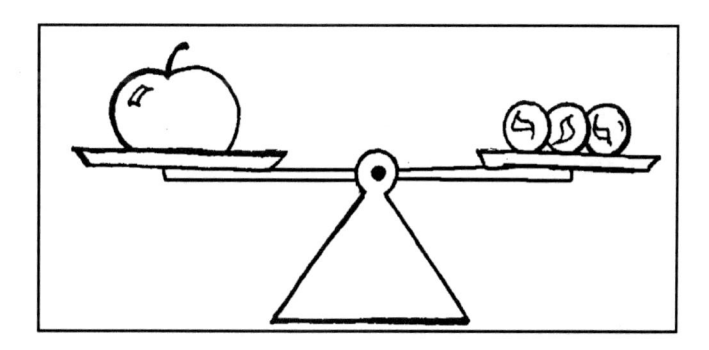

It takes Jennifer 7 minutes to walk to school. How long does it take her to walk there and back?

Making decisions

Lesson 4a

1. Class 3 are measuring the classroom windows. Each window is 2 metres wide, how long would their tape measure have to be to measure 3 windows in a row?

   I will need to _____

   To help me I will use _____

   The answer is _____

2. Jack and Ravi are playing a game. The winner gets a marble. Jack has 3 marbles and Ravi has 5 marbles. How many games have they played so far?

   I will need to _____

   To help me I will use _____

   The answer is _____

3. Sam posted a letter on Monday and his gran received it on Thursday. How many days did it take to get there?

   I will need to _____

   To help me I will use _____

   The answer is _____

4. Rachel's birthday cake weighs the same as 4 oranges. Half of the cake is eaten by Rachel's friends. How many oranges does the rest of the cake weigh?

   I will need to _____

   To help me I will use _____

   The answer is _____

Making decisions

Lesson 4b

1. Class 3 are measuring the classroom windows. Each window is 2 metres wide, how long would their tape measure have to be to measure 4 windows in a row?

   I will need to _____

   To help me I will use _____

   The answer is _____

2. Jack and Ravi are playing a game. The winner gets a marble. Jack has 7 marbles and Ravi has 6 marbles. How many games have they played so far?

   I will need to _____

   To help me I will use _____

   The answer is _____

3. Sam posted a letter on Monday and his gran received it on the next Monday. How many days did it take to get there?

   I will need to _____

   To help me I will use _____

   The answer is _____

4. Rachel's birthday cake weighs the same as 10 oranges. Half of the cake is eaten by Rachel's friends. How many oranges does the rest of the cake weigh?

   I will need to _____

   To help me I will use _____

   The answer is _____

Making decisions

1. Class 3 are measuring the classroom windows. Each window is 2 metres wide, how long would their tape measure have to be to measure 6 windows in a row?

   I will need to _____

   To help me I will use _____

   The answer is _____

2. Jack and Ravi are playing a game. The winner gets a marble. Jack has 11 marbles and Ravi has 6 marbles. How many games have they played so far?

   I will need to _____

   To help me I will use _____

   The answer is _____

3. Sam posted a letter on Monday and his gran received it after the weekend on the next Wednesday. How many days did it take to get there?

   I will need to _____

   To help me I will use _____

   The answer is _____

4. Rachel's birthday cake weighs the same as 16 oranges. Half of the cake is eaten by Rachel's friends. How many oranges does the rest of the cake weigh?

   I will need to _____

   To help me I will use _____

   The answer is _____

Making decisions

Lesson 5

Whole class activity

Leon has a jug that holds 6 cups of squash. How many friends could he give drinks to if he had 2 jugs?

One letter costs 5p to send. One letter costs 8p to send. How much money will Kamal need to send the 2 letters?

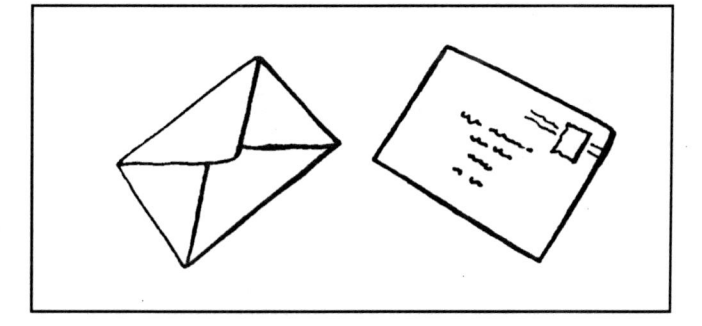

The school library has 11 new books. 5 of them have a hard cover. How many have a soft cover?

Making decisions

Lesson 5a

1. It costs £3 to buy a football. Robert has £6. How much money will he have left if he buys the football?

   I will need to _____

   To help me I will use _____

   The answer is _____

2. Charlotte has a strawberry lace that is 5 rulers long and Tom has a lace that is 2 rulers long. How long would their strawberry laces be if they put them together?

   I will need to _____

   To help me I will use _____

   The answer is _____

3. Anna and Joe are going on a picnic. Anna's drinks bottle holds 1 cup of squash, Joe's holds 4 cups. How many cups of squash do they have all together?

   I will need to _____

   To help me I will use _____

   The answer is _____

4. 4 school chairs weigh the same as a table. Sophie's dad has a van that can carry 2 tables. How many chairs could it carry?

   I will need to _____

   To help me I will use _____

   The answer is _____

Making decisions

Lesson 5b

1. It costs £5 to buy a football. Robert has £10. How much money will he have left if he buys the football?

I will need to _____

To help me I will use _____

The answer is _____

2. Charlotte has a strawberry lace that is 8 rulers long and Tom has a lace that is 6 rulers long. How long would their strawberry laces be if they put them together?

I will need to _____

To help me I will use _____

The answer is _____

3. Anna and Joe are going on a picnic. Anna's drinks bottle holds 3 cups of squash, Joe's holds 5 cups. How many cups of squash do they have all together?

I will need to _____

To help me I will use _____

The answer is _____

4. 6 school chairs weigh the same as a table. Sophie's dad has a van that can carry 2 tables. How many chairs could it carry?

I will need to _____

To help me I will use _____

The answer is _____

Making decisions

1. It costs £7 to buy a football. Robert has £15. How much money will he have left if he buys the football?

I will need to _____

To help me I will use _____

The answer is _____

2. Charlotte has a strawberry lace that is 11 rulers long and Tom has a lace that is 10 rulers long. How long would their strawberry laces be if they put them together?

I will need to _____

To help me I will use _____

The answer is _____

3. Anna and Joe are going on a picnic. Anna's drinks bottle holds 7 cups of squash, Joe's holds 10 cups. How many cups of squash do they have all together?

I will need to _____

To help me I will use _____

The answer is _____

4. 10 school chairs weigh the same as a table. Sophie's dad has a van that can carry 2 tables. How many chairs could it carry?

I will need to _____

To help me I will use _____

The answer is _____

Making decisions

Whole class activity

It takes Richard 9 minutes to run around the school field. It takes Efia 6 minutes more. How long does it take Efia to run around the field?

Sean's fruit cake takes 3 hours to cook. If he puts it in the oven at 1 o'clock what time should he take it out?

Tina has made a tower with blocks. It is 7 blocks high. Keith has made a tower twice as big. How many blocks did he use?

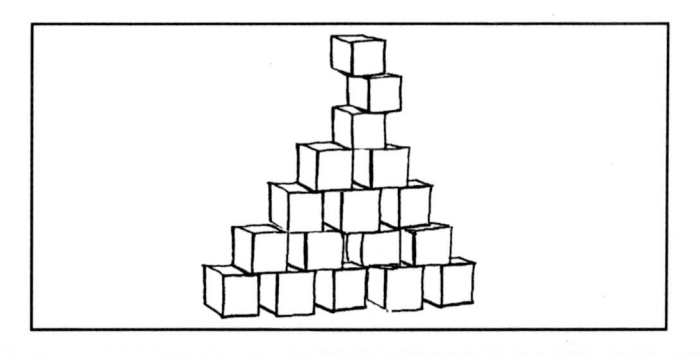

Making decisions

1. When Hinda uses her paddling pool her dad fills it with 8 jugs of water. Afterwards he lets out 3 jugs. How much water is still left in the paddling pool?

   I will need to _____

   To help me I will use _____

   The answer is _____

2. For his birthday Lee is given £4 by his grandma and £4 by his uncle. How much birthday money does he have?

   I will need to _____

   To help me I will use _____

   The answer is _____

3. Paul and Christopher are measuring the length of their feet. Paul's feet are 6 cubes long. Christopher's feet are 3 cubes longer. How many cubes long are Christopher's feet?

   I will need to _____

   To help me I will use _____

   The answer is _____

4. A cup of flour weighs about the same as 5 marbles. Susan's biscuit recipe needs 2 cups of flour. How many marbles would balance two cups of flour?

   I will need to _____

   To help me I will use _____

   The answer is _____

Making decisions

Lesson 6b

1. When Hinda uses her paddling pool her dad fills it with 12 jugs of water. Afterwards he lets out 4 jugs. How much water is still left in the paddling pool?

   I will need to _____

   To help me I will use _____

   The answer is _____

2. For his birthday Lee is given £7 by his grandma and £4 by his uncle. How much birthday money does he have?

   I will need to _____

   To help me I will use _____

   The answer is _____

3. Paul and Christopher are measuring the length of their feet. Paul's feet are 12 cubes long. Christopher's feet are 5 cubes longer. How many cubes long are Christopher's feet?

   I will need to _____

   To help me I will use _____

   The answer is _____

4. A cup of flour weighs about the same as 7 marbles. Susan's biscuit recipe needs 2 cups of flour. How many marbles would balance two cups of flour?

   I will need to _____

   To help me I will use _____

   The answer is _____

This page may be photocopied by the purchasing institution only.
© Catherine Yemm

www.brilliantpublications.co.uk

MathsProblem Solving – Year 1

Lesson 6C — Making decisions

1. When Hinda uses her paddling pool her dad fills it with 15 jugs of water. Afterwards he lets out 5 jugs. How much water is still left in the paddling pool?

   I will need to _____

   To help me I will use _____

   The answer is _____

2. For his birthday Lee is given £8 by his grandma and £10 by his uncle. How much birthday money does he have?

   I will need to _____

   To help me I will use _____

   The answer is _____

3. Paul and Christopher are measuring the length of their feet. Paul's feet are 16 cubes long. Christopher's feet are 5 cubes longer. How many cubes long are Christopher's feet?

   I will need to _____

   To help me I will use _____

   The answer is _____

4. A cup of flour weighs about the same as 11 marbles. Susan's biscuit recipe needs 2 cups of flour. How many marbles would balance two cups of flour?

   I will need to _____

   To help me I will use _____

   The answer is _____

Reasoning about numbers or shapes

Lesson 1

Whole class activity

You have two shaded squares and two spotted squares. How many different ways can you join them together so that no shaded squares touch and no spotted squares touch?

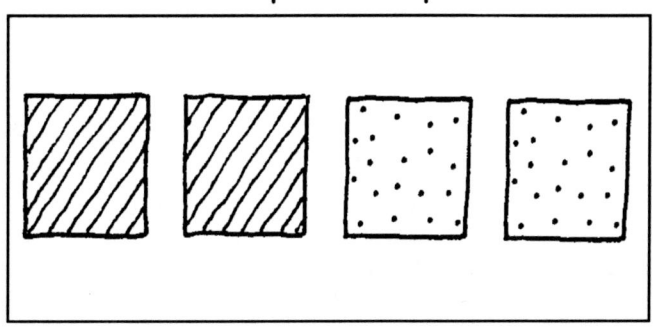

A pentagon has more corners than a square. Circle the correct answer.

True

False

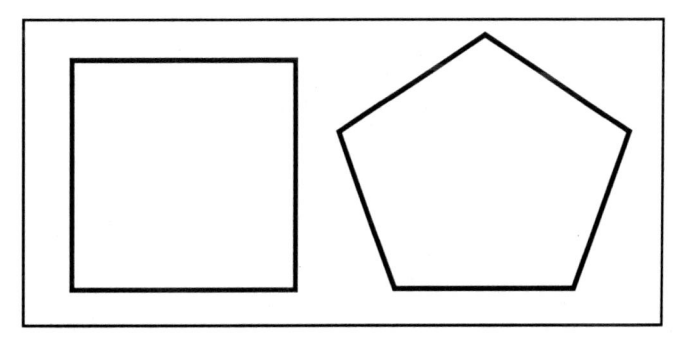

Which has more sides, a triangle or a square?

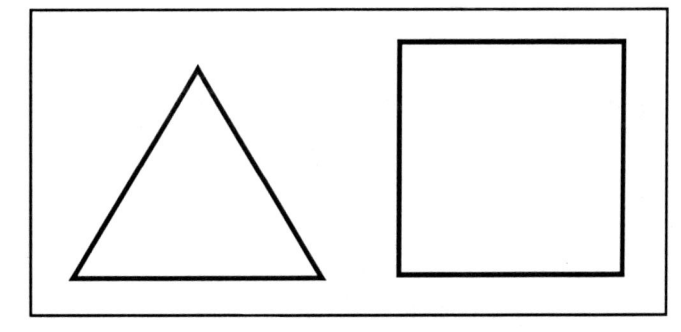

Reasoning about numbers or shapes

Lesson 1a

1. A triangle has more sides than a circle.
   Circle the correct answer.

   True

   False

2. Can triangles be put together to make squares?

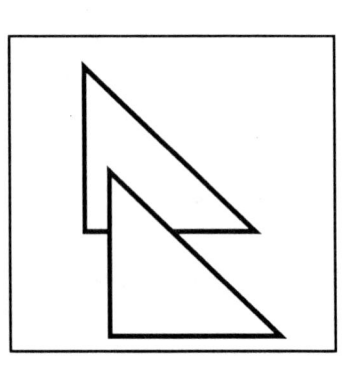

3. How many squares can you count?

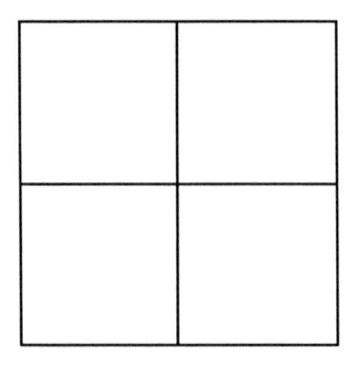

4. Look at this pattern.

   ● ● ■ ▲ ● ● ■ ▲ ● ● ■ ▲ ● ● ■ ▲

   How many times is each shape repeated?

   ●
   ■
   ▲

Reasoning about numbers or shapes

Lesson 1b

1. A rectangle has more sides than a triangle.
   Circle the correct answer.

   True

   False

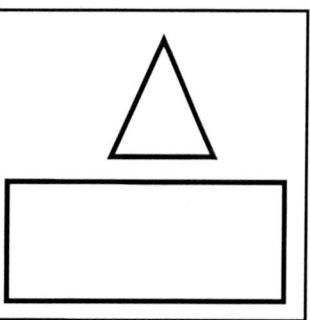

2. Can triangles be put together to make rectangles?

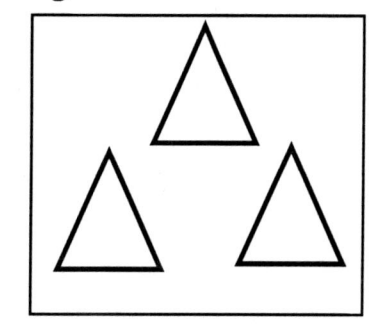

3. How many squares can you count?

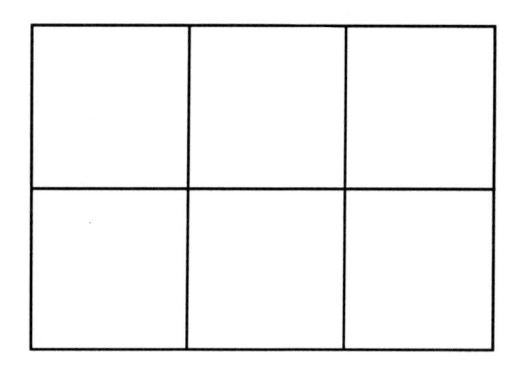

4. Look at this pattern.

   ✚ ● ● ■ ▲ ● ● ■ ✚ ● ● ■ ▲ ● ● ■

   How many times is each shape repeated?

   ●
   ■
   ▲
   ✚

Reasoning about numbers or shapes

Lesson 1C

1. A pentagon has more sides than a rectangle.
   Circle the correct answer.

   True

   False

2. Can triangles be put together to make a shape called a trapezium?

3. How many squares can you count?

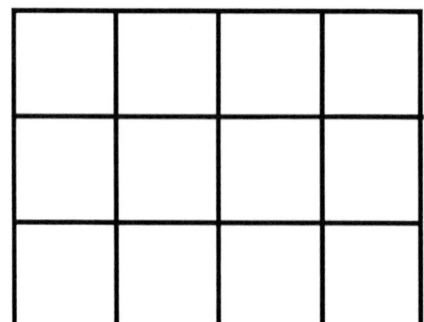

4. Look at this pattern.

   ✚ ✣ ● ✤ ● ■ ▲ ● ✤ ● ■ ✚ ✣ ● ✤ ● ■ ▲ ● ✤ ● ■

   How many times is each shape repeated?

   ●
   ■
   ▲
   ✚
   ✣

www.brilliantpublications.co.uk
Maths Problem Solving – Year 1

Whole class activity

Reasoning about numbers or shapes

Lesson 2

If you add 2 + 2 + 2, is the answer odd or even?

How much money is a 5p coin, a 2p coin and two 1p coins?

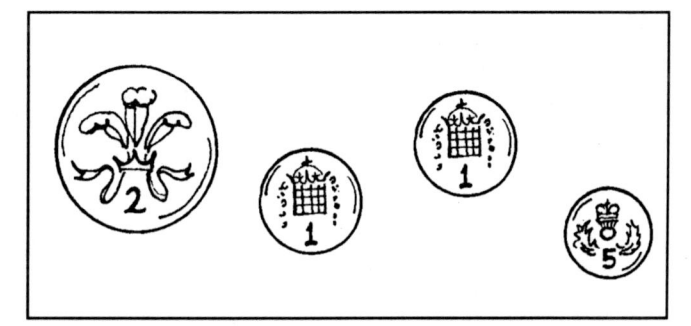

Can you name 4 pairs of numbers with a difference of 3?

Lesson 2a

Reasoning about numbers or shapes

1. You have a bag of 5p, 2p and 1p coins. Can you find three or more ways of making 8p?

2. These cards can be put into pairs so each pair has a difference of 2. Draw a line to link up the pairs.

   | 7 | 3 | 9 | 4 | 5 | 2 |

3. Using the digits 1 and 2, write down a different number in each of these three boxes.

4. Which numbers are missing in this pattern?

   1 2 ___ 4 5 ___ ___ 8 ___ 10

Reasoning about numbers or shapes

Lesson 2b

1. You have a bag of 5p, 2p and 1p coins. Can you find three or more ways of making 10p?

2. These cards can be put into pairs so each pair has a difference of 4. Draw a line to link up the pairs.

| 7 | 3 | 9 | 11 | 5 | 15 |

3. Using the digits 1 and 2, write down a different number in each of these four boxes.

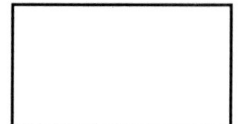

4. Which numbers are missing in this pattern?

0   2   ___   6   8   ____   ____   14   ____   18

Reasoning about numbers or shapes

Lesson 2c

1. You have a bag of 5p, 2p and 1p coins. Can you find three or more ways of making 15p?

2. These cards can be put into pairs so each pair has a difference of 5. Draw a line to link up the pairs.

   | 0 | 15 | 5 | 11 | 6 | 20 |

3. Using the digits 1 and 2, write down a different number in each of these six boxes.

4. Which numbers are missing in this pattern?

   1   4   ___   10   13   ___   ___   22   ___   28

Whole class activity

Reasoning about numbers or shapes

Lesson 3

Make up a pattern with these shapes that repeats 3 times.

Can you put 5 squares together to make a cross?

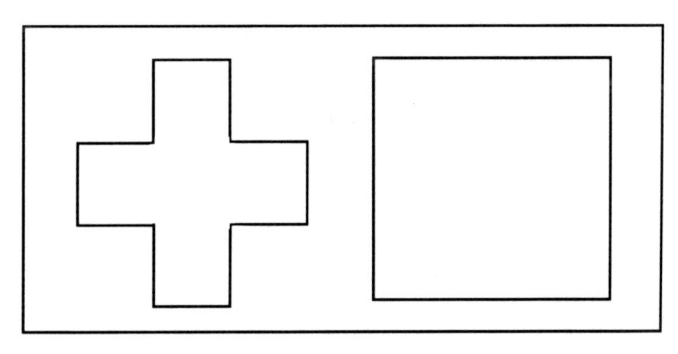

Can you name two 3D shapes which have 6 faces?

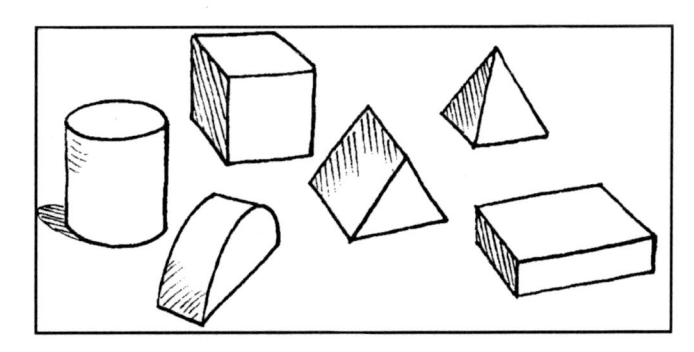

Reasoning about numbers or shapes

Lesson 3a

A cylinder has more edges than a cone.

1. True

   False

2. This is the number 1

   Can you make the number 3 out of squares?

3. Repeat this pattern three times. What will be the 5th shape?

   ○ □

4. Can you name any 3D shapes which have no corners?

Reasoning about numbers or shapes

Lesson 3b

1. A cube has more edges than a cylinder.

   True

   False

2. This is the number 1

   Can you make the number 13 out of squares?

3. Repeat this pattern three times. What will be the 8th shape?

   ○ □ ○

4. Can you name two 3D shapes that have eight corners?

Reasoning about numbers or shapes

Lesson 3C

A cuboid has more edges than a pyramid.

1. True

   False

2. This is the number 1

   Can you make the number 23 out of squares?

3. Repeat this pattern four times. What will be the 10th shape?

   □ ○ □

4. Can you name three 3D shapes which have more than four corners?

Reasoning about numbers or shapes

Lesson 4

Whole class activity

Can you make four different numbers with the digits 2 and 5?

$$2 \quad 5$$

Add 4 to all of these numbers.

7	3
1	8

How many answers are more than 10?

Which numbers between 3 and 11 are even?

Reasoning about numbers or shapes

1. If you double the odd numbers less than 5 you get an even number.

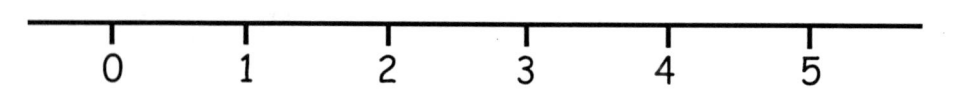

```
0   1   2   3   4   5
```

True

False

2. The number added on each time in this pattern is between 1 and 4. Can you work it out?

1   3   5   7   9   11   13

3. How many different ways can you make 6 from two numbers?

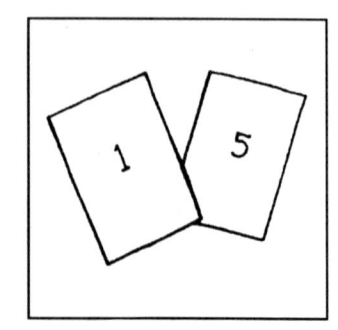

1   5

4. Take 2 from all of these numbers.

10   4   8

Are the answers more or less than 5?

Reasoning about numbers or shapes

Lesson 4b

1. If you double the odd numbers less than 8 you get an even number.

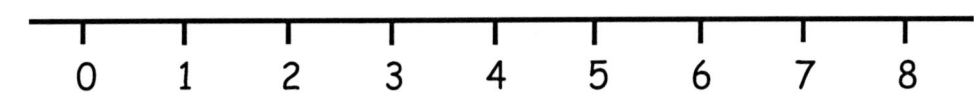

   ```
   |---|---|---|---|---|---|---|---|---|
   0   1   2   3   4   5   6   7   8
   ```

   True

   False

2. The number added on each time in this pattern is between 2 and 5. Can you work it out?

   1   4   7   10   13   16   19

3. How many different ways can you make 8 from two numbers?

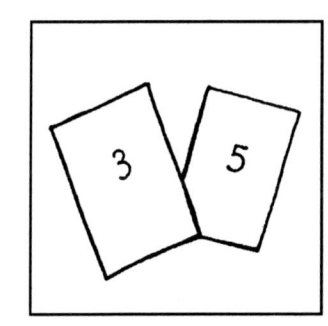

   [cards showing 3 and 5]

4. Take 3 from all of these numbers.

   12   6   9

   Are the answers more or less than 5 ?

Lesson 4C

Reasoning about numbers or shapes

1. If you double the odd numbers less than 12 you get an even number.

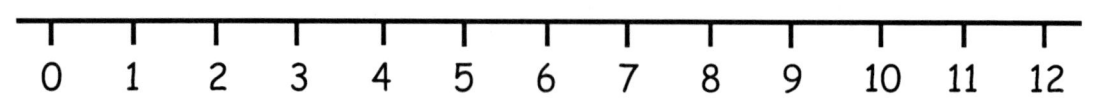

   ```
   ┬───┬───┬───┬───┬───┬───┬───┬───┬───┬───┬───┬───┬
   0   1   2   3   4   5   6   7   8   9   10  11  12
   ```

   True

   False

2. The number added on each time in this pattern is between 3 and 8. Can you work it out?

   1   6   11   16   21   26

3. How many different ways can you make 10 from two numbers?

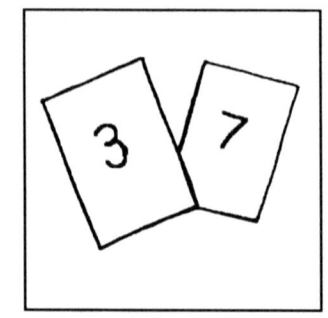

   [cards showing 3 and 7]

4. Take 3 from all of these numbers.

   11   20   15

   Are the answers more or less than 10?

Reasoning about numbers or shapes

Lesson 5

Whole class activity

Join two triangles together. How many sides does your new shape have?

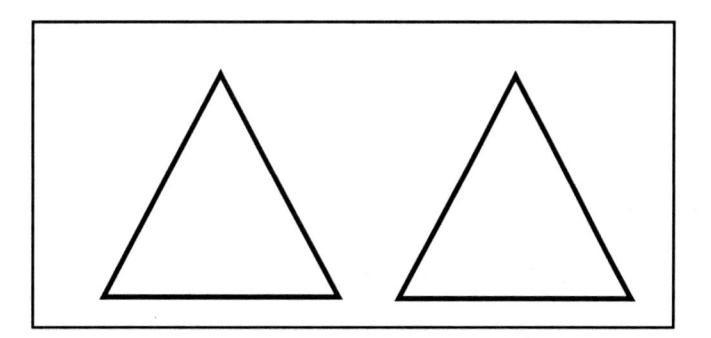

Draw a pattern with three squares and four triangles. Make it repeat three times.

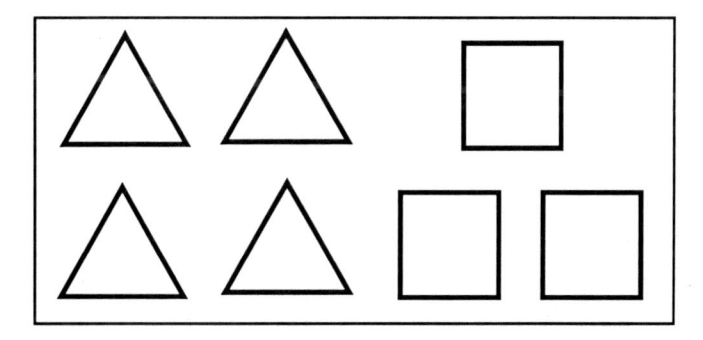

How could you sort these shapes?

Reasoning about numbers or shapes

Lesson 5a

You can make a large triangle with two small triangles.

1. True

   False

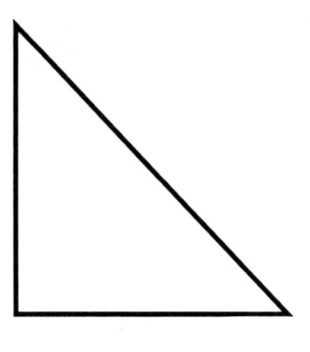

2. There are three shapes that have curved faces.

   True

   False

3. Which shape goes in which box?

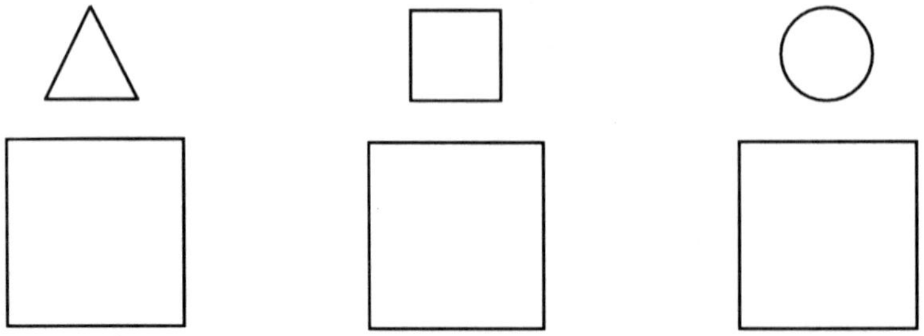

   1 side   3 sides   4 sides

4. How many different ways can you draw a row of two circles? Fill in the grids.

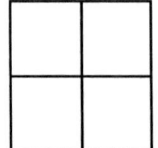

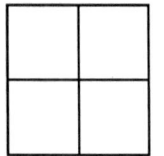

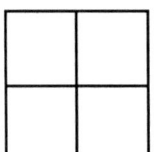

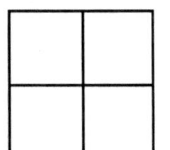

Reasoning about numbers or shapes

Lesson 5b

1. You can make a large triangle with four small triangles.

   True

   False

2. There are two shapes that have curved and flat faces.

   True

   False

3. Which shapes go in which box?

   1 side   3 sides   4 sides

4. How many different ways can you draw a row of two circles? Fill in the grids.

Reasoning about numbers or shapes

Lesson 5c

You can make a large triangle with 16 small triangles.

1. True

   False

2. There are more than three shapes that have flat faces.

   True

   False

3. Which shapes go in which box?

   1 side   3 sides   4 sides   more than 4 sides

4. How many different ways can you draw a row of three circles? Fill in the grids.

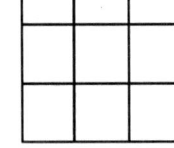

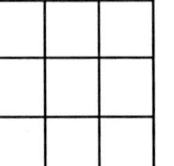

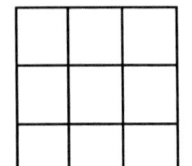

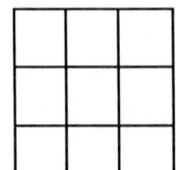

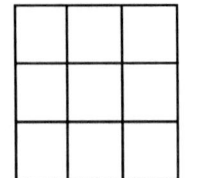

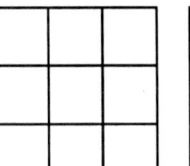

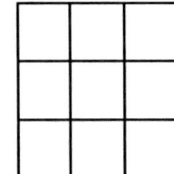

Reasoning about numbers or shapes

Lesson 6

Whole class activity

Take away 2 from each number. Which answers are even?

8　　5　　11　　3

Think of a number and add 7. Is your answer more or less than 12?

Start with the number 3. Add 2, five times. What pattern do you make?

Lesson 6a

Reasoning about numbers or shapes

1. If I add any two of these numbers together the answer will always be more than 3.

    5   3   6   2   4   2

    True

    False

2. If you take an odd number away from 5 you get an even number.

    True

    False

    Show your workings.

3. George is older than 4 but younger than 10. How old could he be?

4. Look at these numbers.

    0 2 3 2 4 1

    Put two numbers in each box so that the numbers in each box add up to 4.

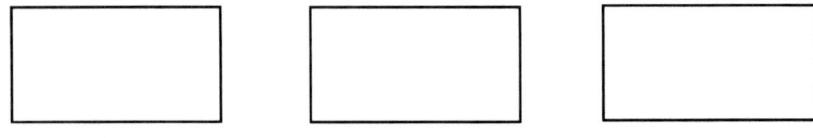

Reasoning about numbers or shapes

Lesson 6b

1. If I add any two of these numbers together the answer will always be more than 5.

         3   7   3   6   4   5

   True

   False

2. If you take an odd number away from 7 you get an even number.

   True

   False

   Show your workings.

3. George is older than 3 but younger than 12. How old could he be?

4. Look at these numbers.

       1 2 3 3 4 5

   Put two numbers in each box so that the numbers in each box add up to 6.

Reasoning about numbers or shapes

Lesson 6C

1. If I add any two of these numbers together the answer will always be more than 7.

    4   8   4   7   5   6

    True

    False

2. If you take an odd number away from 9 you get an even number.

    True

    False

    Show your workings.

3. George is older than 4 but younger than 14. How old could he be?

4. Look at these numbers.

    7   4   2   5   6   4   3   1

    Put two numbers in each box so that the numbers in each box add up to 8.

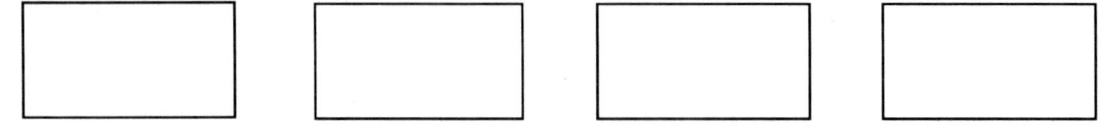

    ☐   ☐   ☐   ☐

Problems involving 'real life', money or measures

Lesson 1

Whole class activity

Jack has 15p and he buys an apple for 6p. How much money does he have left?

Leanne has two strips of paper that are both 8 hands long. If she joins them together how long will the strip be?

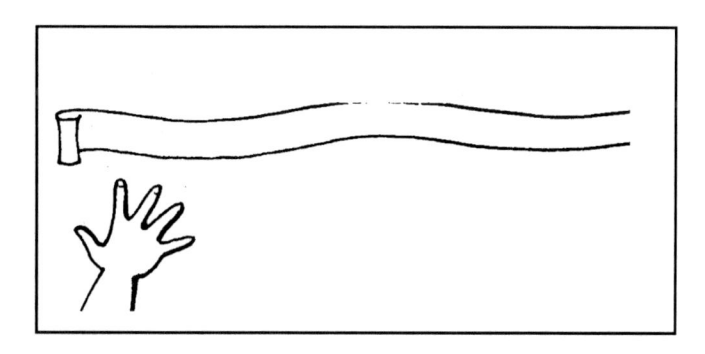

The school bus has 3 passengers. It picks up 4 more children then 2 get off. How many children are on the bus?

Lesson 1a

Problems involving 'real life', money or measures

1. For school dinners on Friday the children can choose a jacket potato or salad. If 8 children are having school dinners and 4 choose a jacket potato, how many children are having salad?

2. Lisa bought a carton of orange squash from the corner shop for 3p. She gave the shopkeeper 5p. How much change did she get?

3. The sports field is 8 metres long. On sports day Dewey runs halfway then falls over. How far did he run?

4. In her money box Kanya has a 5p coin, a 2p coin and a 1p coin. How much money does she have all together?

Problems involving 'real life', money or measures

Lesson 1b

1. For school dinners on Friday the children can choose a jacket potato or salad. If 10 children are having school dinners and 5 choose a jacket potato, how many children are having salad?

2. Lisa bought a carton of orange squash from the corner shop for 6p. She gave the shopkeeper 10p. How much change did she get?

3. The sports field is 12 metres long. On sports day Dewey runs halfway then falls over. How far did he run?

4. In her money box Kanya has a 10p coin, a 2p coin and a 1p coin. How much money does she have all together?

Problems involving 'real life', money or measures

Lesson 1C

1. For school dinners on Friday the children can choose a jacket potato or salad. If 12 children are having school dinners and 7 choose a jacket potato, how many children are having salad?

2. Lisa bought a carton of orange squash from the corner shop for 9p. She gave the shopkeeper 15p. How much change did she get?

3. The sports field is 18 metres long. On sports day Dewey runs halfway then falls over. How far did he run?

4. In her money box Kanya has a 10p coin, a 5p coin, a 2p coin and a 1p coin. How much money does she have all together?

www.brilliantpublications.co.uk

This page may be photocopied by the purchasing institution only.

62 Maths Problem Solving – Year 1

© Catherine Yemm

Whole class activity

Problems involving 'real life', money or measures

Lesson 2

Jaya buys a box of cream eggs. There are 3 eggs in a box. How many eggs are in 2 boxes?

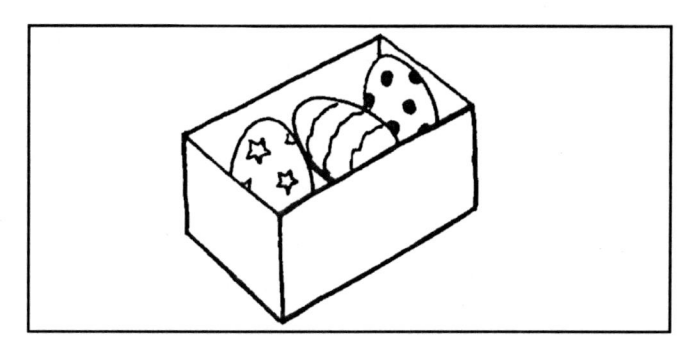

Joshua has read 8 pages of his book. If he reads 7 more how many will he have read?

Tegan starts watching television at 4 o'clock. She watches it for an hour then she has her tea for an hour. What time does she finish her tea?

Lesson 2a

Problems involving 'real life', money or measures

1. I think of a number and add 2. The answer is 10. What was my number?

2. Stacey buys a toy for 9 pence. What coins could she use to pay for her toy?

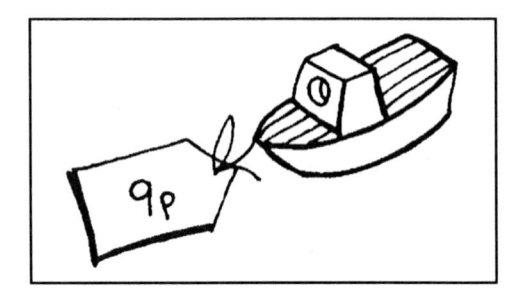

3. Kedar plays a game on the computer. It takes him 6 minutes to find the treasure. It takes his friend Sam half that time. How long does it take Sam to find the treasure?

4. It is 3.00 p.m. Fiona's swimming lesson starts in 3 hours. What time does her lesson start?

Problems involving 'real life', money or measures

Lesson 2b

1. I think of a number and add 5. The answer is 14. What was my number?

2. Stacey buys a toy for 12 pence. What coins could she use to pay for her toy?

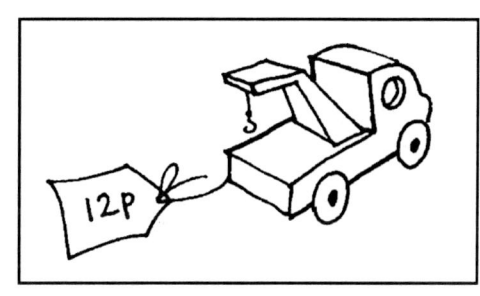

3. Kedar plays a game on the computer. It takes him 14 minutes to find the treasure. It takes his friend Sam half that time. How long does it take Sam to find the treasure?

4. It is 1.00 p.m. Fiona's swimming lesson starts in 4 hours. What time does her lesson start?

Problems involving 'real life', money or measures

Lesson 2C

1. I think of a number and add 9. The answer is 20. What was my number?

2. Stacey buys a toy for 19 pence. What coins could she use to pay for her toy?

3. Kedar plays a game on the computer. It takes him 20 minutes to find the treasure. It takes his friend Sam half that time. How long does it take Sam to find the treasure.

4. It is 11.00 a.m. Fiona's swimming lesson starts in 5 hours. What time does her lesson start?

Problems involving 'real life', money or measures

Lesson 3

Whole class activity

Kane weighs the same as 4 chairs. Lee weighs the same as 8 chairs. How much heavier is Lee?

Lana picks up some cherries. They come in pairs. She picks up 6 pairs. How many cherries does she have?

Sebastian has 12p in his hand. He drops 8p. How much does he have left? What coins could he have dropped?

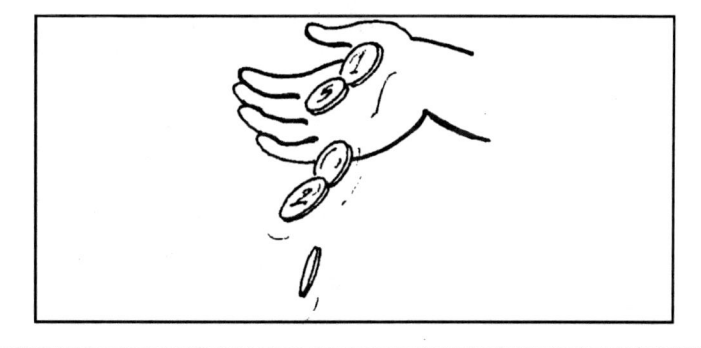

Problems involving 'real life', money or measures

1. At the park 2 children are on the swings, 2 children are on the slide and 3 children are on the roundabout. How many children are playing in the park all together?

2. A ride at the fair is 7p a go. How much would it cost for two goes?

3. Joshua has 12 books on his bedroom shelf. He lends 2 to his friend Kasi and takes 2 back to the library. How many books does he have on his shelf now?

4. John weighs the same as 6 bricks, Lucy weighs the same as 4 bricks. How much heavier is John than Lucy?

Problems involving 'real life', money or measures

Lesson 3b

1. At the park 3 children are on the swings, 4 children are on the slide and 5 children are on the roundabout. How many children are playing in the park all together?

2. A ride at the fair is 9p a go. How much would it cost for two goes?

3. Joshua has 16 books on his bedroom shelf. He lends 2 to his friend Kasi and takes 3 back to the library. How many books does he have on his shelf now?

4. John weighs the same as 9 bricks, Lucy weighs the same as 4 bricks. How much heavier is John than Lucy?

Problems involving 'real life', money or measures

Lesson 3C

1. At the park 4 children are on the swings, 6 children are on the slide and 3 children are on the roundabout. How many children are playing in the park all together?

2. A ride at the fair is 11p a go. How much would it cost for two goes?

3. Joshua has 20 books on his bedroom shelf. He lends 4 to his friend Kasi and takes 6 back to the library. How many books does he have on his shelf now?

4. John weighs the same as 18 bricks, Lucy weighs the same as 11 bricks. How much heavier is John than Lucy?

www.brilliantpublications.co.uk
70 Maths Problem Solving – Year 1

This page may be photocopied by the purchasing institution only.
© Catherine Yemm

Problems involving 'real life', money or measures

Lesson 4

Whole class activity

There are 10 scissors in the blue pot and 14 scissors in the red pot. How many scissors are there in total?

Sheba can skip for 2 minutes. Brady can skip for 5 minutes longer. How long can Brady skip for?

Lucy is 2 hands taller than Ruth. Ruth is 16 hands tall. How tall is Lucy?

Problems involving 'real life', money or measures

Lesson 4a

1. An apple costs 5p, a banana costs 4p and an orange costs 5p. You have a 10p coin. Which two pieces of fruit could you buy?

2. Rachel and Mark are playing a board game with a spinner. Mark spins a 6 and moves his counter but then he has to go back 2 spaces. How far has he moved?

3. Jamie's mum has bought a new car. It is 7 feet long. Their caravan is 3 feet longer. How long is their caravan?

4. A pack of stickers cost 10p but they are in the half price sale. How much are they now?

Problems involving 'real life', money or measures

Lesson 4b

1. An apple costs 7p, a banana costs 7p and an orange costs 8p. You have a 10p and a 5p coin. Which two pieces of fruit could you buy?

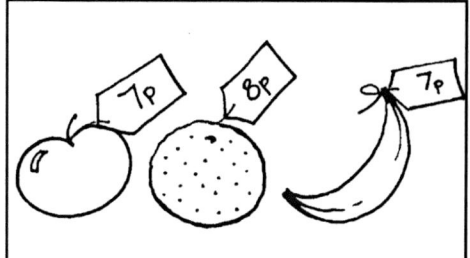

2. Rachel and Mark are playing a board game with a spinner. Mark spins a 10 and moves his counter but then he has to go back 4 spaces. How far has he moved?

3. Jamie's mum has bought a new car. It is 8 feet long. Their caravan is 7 feet longer. How long is their caravan?

4. A pack of stickers cost 16p but they are in the half price sale. How much are they now?

Lesson 4C

Problems involving 'real life', money or measures

1. An apple costs 12p, a banana costs 9p and an orange costs 7p. You have a 20p coin. Which two pieces of fruit could you buy?

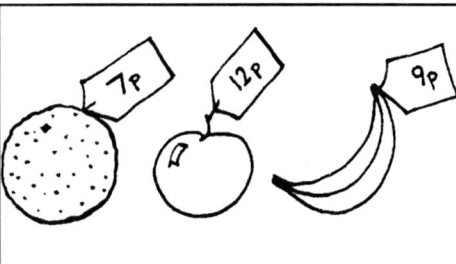

2. Rachel and Mark are playing a board game with a spinner. Mark spins a 14 and moves his counter but then he has to go back 5 spaces. How far has he moved?

3. Jamie's mum has bought a new car. It is 9 feet long. Their caravan is 8 feet longer. How long is their caravan?

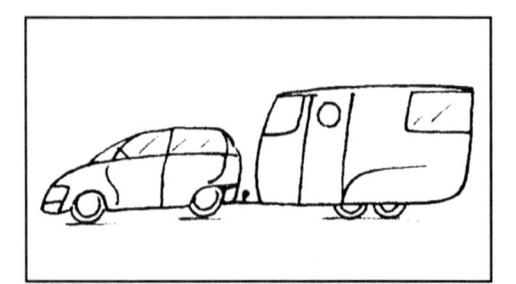

4. A pack of stickers cost 18p but they are in the half price sale. How much are they now?

Problems involving 'real life', money or measures

Lesson 5

Whole class activity

Between them, Gavin and Shane, have twice as many sweets as Paula. If Paula has 4 sweets, how many do they have all together?

Genevieve, Tom and Charlotte have three hamsters each. How many do they have all together?

Adrian scored 13 tries in his first game of rugby but 4 fewer in his second game. How many did he score in his second game?

Problems involving 'real life', money or measures

Lesson 5a

1. Lucy has 1 mouse and 2 cats. How many tails are there in her house all together?

2. Karen has 5 euros pocket money each week. If she saves it all, how much money will she have after two weeks?

3. It takes Ben 2 minutes to walk to school and another 4 minutes to walk from the school to the park. How long would it take him to walk straight from his house to the park?

4. Oliver invites 10 friends to his birthday party. 2 friends cannot come as they are on holiday and 2 friends are ill. How many friends will be at Oliver's party?

Problems involving 'real life', money or measures

Lesson 5b

1. Lucy has 2 mice and 3 cats. How many tails are there in her house all together?

2. Karen has 5 euros pocket money each week. If she saves it all, how much money will she have after three weeks?

3. It takes Ben 6 minutes to walk to school and another 5 minutes to walk from the school to the park. How long would it take him to walk straight from his house to the park?

4. Oliver invites 13 friends to his birthday party. 3 friends cannot come as they are on holiday and 2 friends are ill. How many friends will be at Oliver's party?

Problems involving 'real life', money or measures

Lesson 5c

1. Lucy has 5 mice, 3 cats, 2 dogs and 4 gerbils. How many tails are there in her house all together?

2. Karen has 5 euros pocket money each week. If she saves it all, how much money will she have after four weeks?

3. It takes Ben 8 minutes to walk to school and another 6 minutes to walk from the school to the park. How long would it take him to walk straight from his house to the park?

4. Oliver invites 17 friends to his birthday party. 4 friends cannot come as they are on holiday and 2 friends are ill. How many friends will be at Oliver's party?

Problems involving 'real life', money or measures

Lesson 6

Whole class activity

Brenda spent £7 on a train ticket and £4 on a bus ticket. How much change did she have from £15?

A cereal bar is 8p. How much would it cost to buy three cereal bars?

Owen gets up at 8 o'clock on Monday. Debra gets up two hours earlier. What time does Debra get up?

Problems involving 'real life', money or measures

1. It is pancake day. Ramya needs 3 cups of milk to make a batch of pancakes. How many cups of milk will she need to make 2 batches of pancakes?

2. Philip buys 5 sweets at 2p each. What coins could he use to pay for them?

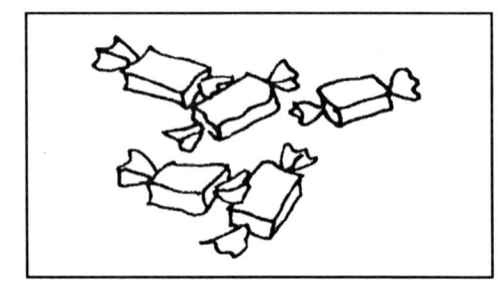

3. James is given £8 for his birthday. He buys a football for £2 and a book for £1. How much money does he have left?

4. Sarah has got 8 Christmas cards in her bag. She gives half of them away. How many has she got left?

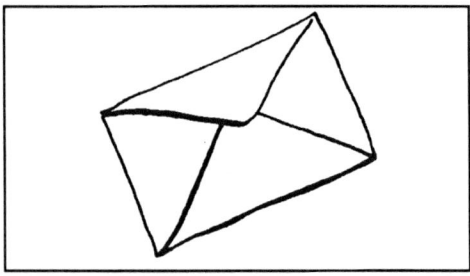

Problems involving 'real life', money or measures

Lesson 6b

1. It is pancake day. Ramya needs 3 cups of milk to make a batch of pancakes. How many cups of milk will she need to make 4 batches of pancakes?

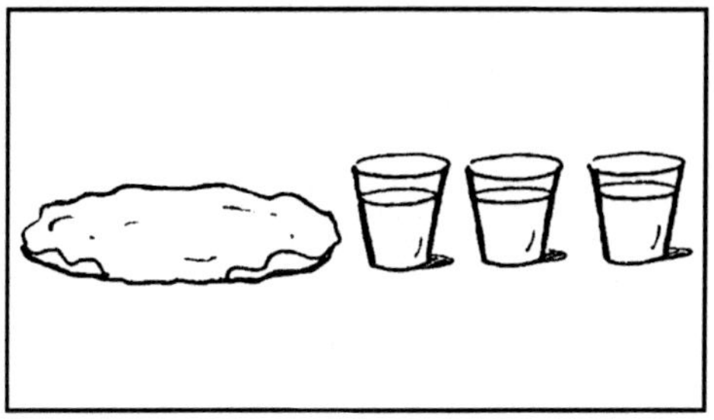

2. Philip buys 7 sweets at 2p each. What coins could he use to pay for them?

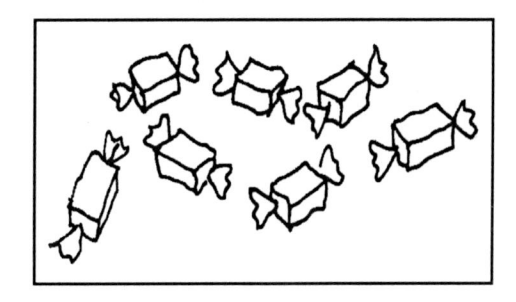

3. James is given £12 for his birthday. He buys a football for £3 and a book for £2. How much money does he have left?

4. Sarah has got 16 Christmas cards in her bag. She gives half of them away. How many has she got left?

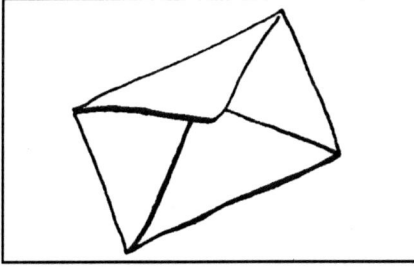

Problems involving 'real life', money or measures

Lesson 6C

1. It is pancake day. Ramya needs 4 cups of milk to make a batch of pancakes. How many cups of milk will she need to make 4 batches of pancakes?

2. Philip buys 10 sweets at 2p each. What coins could he use to pay for them?

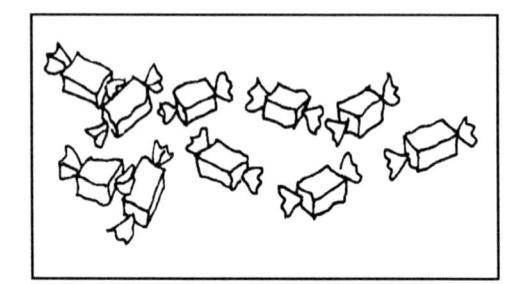

3. James is given £15 for his birthday. He buys a football for £2 and a book for £3. How much money does he have left?

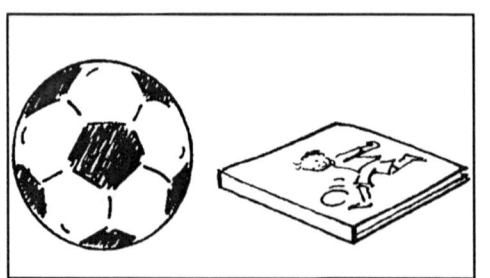

4. Sarah has got 22 Christmas cards in her bag. She gives half of them away. How many has she got left?

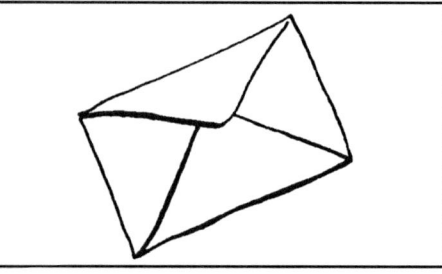

Organizing and using data

Lesson 1

Whole class activity

The school caretaker wants some children to help him pick up stones from the playground. He wants children who can pick up lots of stones at the same time so that the job gets done quickly.

The children in the class are called Joe, Samantha, Rati, Fraser, Emily and Karl.

Write down the names of the children as a list.

1. _____   4. _____
2. _____   5. _____
3. _____   6. _____

The children each practise picking up stones. Emily can hold 3 stones in her hand while Joe can hold 7. Rati and Fraser can both hold 5 while Samantha can hold only 2. Karl can hold 4.

Can you put this information into a table so that we can see it more clearly?

Name	Number of stones

Who can hold the most stones? _____
Who can hold the least? _____
Who do you think has the smallest hands? _____
Why? _____
Which three children should the caretaker choose to help him?
_____

Lesson 1a — Organizing and using data

The school canteen has decided to sell fruit after lunch but they cannot decide which fruit they should sell. Make a list of three fruits which would be good to sell at lunch time.

1. _____
2. _____
3. _____

Ask 6 children in your class which one of the fruits they would choose to eat after their lunch. Put their answers into this table. You could write the name of the fruit or you could draw a picture.

Name	Fruit

Count how many children want each type of fruit. Put your answers in this table.

Type of fruit	Number of children who like it

Which fruit do you think the school should sell after lunch?

_____

Why? _____

_____

Organizing and using data

Lesson 1b

The school canteen has decided to sell fruit after lunch but they cannot decide which fruit they should sell. Make a list of four fruits which would be good to sell at lunch time.

1. _____
2. _____
3. _____
4. _____

Ask 8 children in your class which one of the fruits they would choose to eat after their lunch. Put their answers into this table. You could write the name of the fruit or you could draw a picture.

Name	Fruit	Name	Fruit

Count how many children want each type of fruit. Put your answers in this table.

Type of fruit	Number of children who like it

Which fruit should the school sell after lunch? _____

Why? _____

Is there a fruit that you think would not sell very well? _____

_____

Why? _____

Organizing and using data

Lesson 1C

The school canteen has decided to sell fruit after lunch but they cannot decide which fruit they should sell. Make a list of five fruits which would be good to sell at lunch time.

1. _____
2. _____
3. _____
4. _____
5. _____

Ask 10 children in your class which one of the fruits they would choose to eat after their lunch. Put their answers into this table. You could write the name of the fruit or you could draw a picture.

Name	Fruit	Name	Fruit

Count how many children want each type of fruit. Put your answers in this table.

Type of fruit	Number of children who like it

Which fruit do you think the school should sell after lunch? _____ Why? _____

Is there a fruit that you think would not sell very well? _____ Why? _____

Which fruit would you like? _____

Add your name to the list. How many like that fruit now? _____

Organizing and using data

Lesson 2

Whole class activity

The children in class 1 have been looking at the flowers that grow in the school field.

This is a list of the children and the flowers they found.

Name	Flowers		
Leanne	2 daisies	1 buttercup	0 bluebells
Jimmy	1 daisy	1 buttercup	0 bluebells
Kayleigh	1 daisy	2 buttercups	1 bluebells
Ruth	0 daisies	1 buttercup	0 bluebells
Monica	0 daisies	2 buttercups	1 bluebell
Thaman	2 daisies	2 buttercups	2 bluebells

Can you organize the information in a different way?

Flowers	Number of flowers found
daisies	
buttercups	
bluebells	

Can you build a block graph with your results?

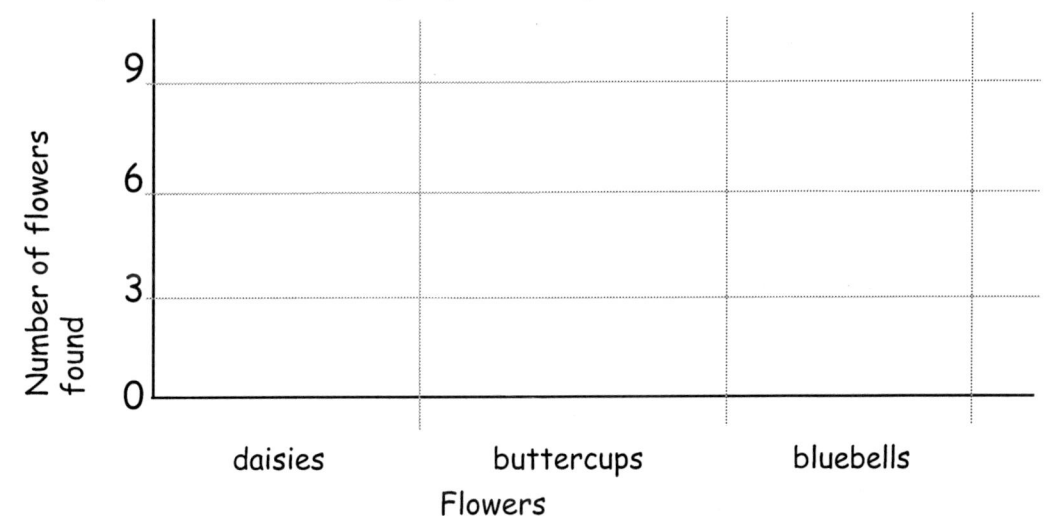

1. Which flower grows in the biggest numbers in the field?

2. Which flower grows in the smallest numbers in the field?

© Catherine Yemm

This page may be photocopied by the purchasing institution only.

www.brilliantpublications.co.uk

Maths Problem Solving – Year 1   87

Organizing and using data

Lesson 2a

Mrs Harrison has decided that her classroom needs a coat of paint but she cannot decide on the colour. She has decided to let the children choose the colour, so she needs to find out what they like.

This is a list of the children in her class and their favourite colour.

Name	Colour
Lucy	Blue
Jacob	Yellow
Bethany	Red
Richard	Green
Mela	Blue
Tamsin	Blue

Can you organize the information in a different way?

Colour	Number of children who like it
Yellow	
Green	
Red	
Blue	

Can you build a block graph with your results?

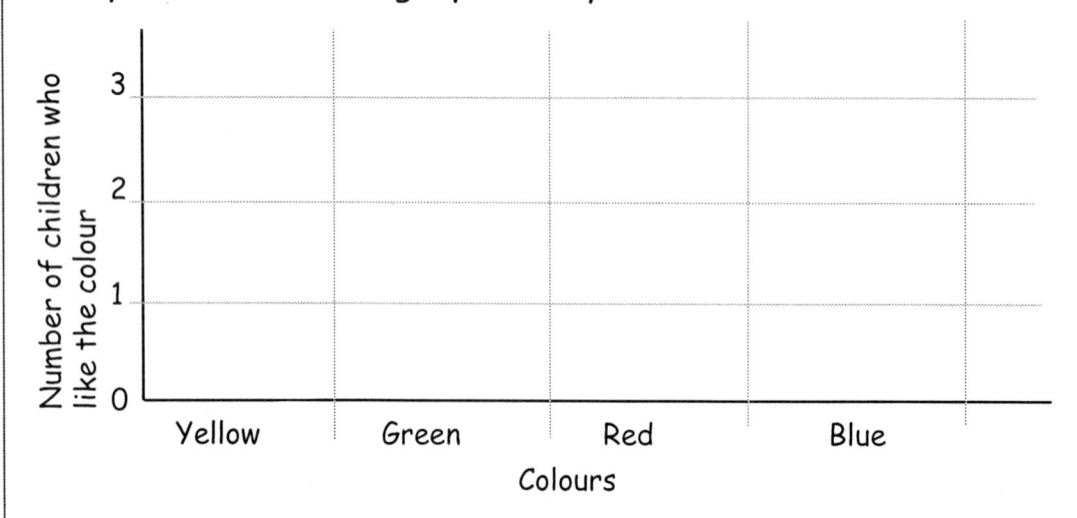

(Block graph: y-axis "Number of children who like the colour" from 0 to 3; x-axis "Colours" with Yellow, Green, Red, Blue)

Which colour do you think Mrs Harrison should paint the classroom? _____
Which colour do you like? _____

Organizing and using data

Lesson 2b

Mrs Harrison has decided that her classroom needs a coat of paint but she cannot decide on the colour. She has decided to let the children choose the colour, so she needs to find out what they like.

This is a list of the children in her class and their favourite colour.

Name	Colour	Name	Colour
Lucy	Blue	Tamsin	Blue
Jacob	Yellow	Hamish	Yellow
Bethany	Red	Kate	Yellow
Richard	Green	Ethan	Blue
Mela	Blue	Harry	Red

Can you organize the information in a different way?

Colour	Number of children who like it
Yellow	
Green	
Red	
Blue	

Can you build a block graph with your results?

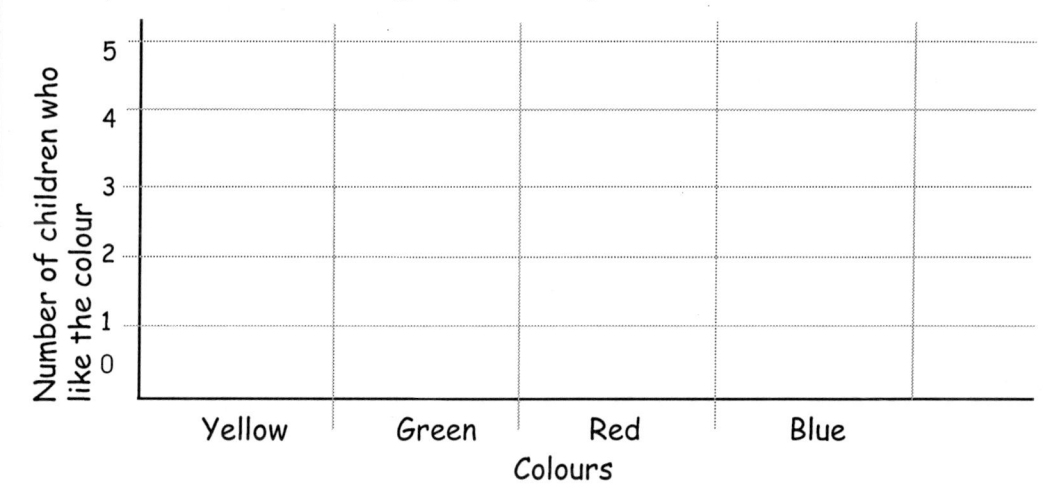

Which colour do most children like? _____
Which colour do children like least? _____
Which colour do you like? _____

Lesson 2C

Organizing and using data

Mrs Harrison has decided that her class needs a coat of paint but she cannot decide on the colour. She has decided to let the children choose the colour, so she needs to find out what they like.

This is a list of the children in her class and their favourite colour.

Name	Colour	Name	Colour
Lucy	Blue	Ethan	Red
Jacob	Yellow	Harry	Yellow
Bethany	Red	Shannon	Red
Richard	Yellow	Tiffany	Green
Mela	Blue	Michael	Blue
Tamsin	Green	Ramya	Blue
Hamish	Red	George	Yellow
Kate	Yellow	Susan	Yellow

Can you organize the information in a different way?

Colour	Number of children who like it
Yellow	
Green	
Red	
Blue	

Can you build a block graph with your results?

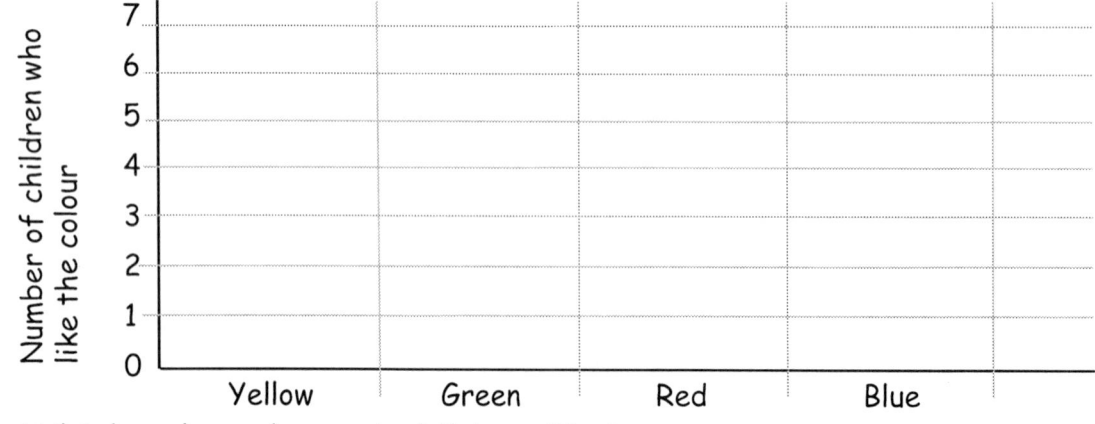

[Block graph with y-axis "Number of children who like the colour" labeled 0–7, and x-axis categories: Yellow, Green, Red, Blue]

Which colour do most children like? _____
Which colour do children like least? _____
How many more children like yellow than blue? _____
Which colour do you like? _____

Whole class activity

Organizing and using data

Lesson 3

The local vet is doing a survey to find out what pets the children in the village own.

The children in the class are called Roma, Jeffrey, Jade, Bethany, Tim and Kirsten. Write down the names of the children in a list.

1. _____
2. _____
3. _____
4. _____
5. _____
6. _____

All of the children have pets. Roma has 2 goldfish and a hamster. Jeffrey has 4 dogs. Jade has a parrot, a chinchilla and 2 stick insects. Bethany has 6 angel fish and 2 terrapins. Tim and Kirsten both have 3 cats and 4 gerbils.

Can you put this information into a table so that we can see it more clearly?

Name	Number of pets

Now answer these questions.

Who has the most pets? _____

Who has the least? _____

Who has more than two pets? _____

How many pets do the children have between them? _____

Organizing and using data

Lesson 3a

Mr Jones would like to start football practice after school but he cannot decide which night to have it on. He decides to see which night most children in his class are free.

Mr Jones asks 7 children in his class when they are free after school. This is what they say.

Debra	I am free on Monday and Tuesday
Charlie	I am free on Monday and Thursday and Wednesday
Jafar	I am free on Friday and Tuesday
Rebecca	I am free on Friday and Tuesday
Chloe	I am free on Monday and Tuesday
Amber	I am free on Monday and Friday
Tim	I am free on Thursday and Tuesday

Look at the days the children are free. Put your information in this table.

Day	Number of children who are free
Monday	
Tuesday	
Wednesday	
Thursday	
Friday	

Which day should Mr Jones have his football lesson on?
_____

Why? _____

Which days would you be able to go on? _____
_____

Organizing and using data

Lesson 3b

Mr Jones would like to start football practice after school but he cannot decide which night to have it on. He decides to see which night most children in his class are free.

Mr Jones asks 10 children in his class when they are free after school. This is what they say.

Debra	I am free on Monday and Tuesday
Charlie	I am free on Monday and Thursday and Wednesday
Jafar	I am free on Friday and Tuesday
Rebecca	I am free on Friday and Tuesday
Chloe	I am free on Monday and Tuesday
Amber	I am free on Monday and Friday
Tim	I am free on Tuesday and Thursday
Sophie	I am free on Monday and Tuesday
Zoë	I am free on Friday
Robert	I am free on Tuesday and Wednesday

Look at the days the children are free. Put your information in this table.

Day	Number of children who are free
Monday	
Tuesday	
Wednesday	
Thursday	
Friday	

Which day are most children free? _____

Which day are fewest children free? _____

Which days would you be able to go on? _____

Organizing and using data

Lesson 3C

Mr Jones would like to start football practice after school but he cannot decide which night to have it on. He decides to see which night more children in his class are free.

Mr Jones asks 12 children in his class when they are free after school. This is what they say.

Debra	I am free on Monday and Tuesday
Charlie	I am free on Monday and Thursday and Wednesday
Jafar	I am free on Friday and Tuesday
Rebecca	I am free on Friday and Tuesday
Chloe	I am free on Monday and Tuesday
Amber	I am free on Monday and Friday
Tim	I am free on Thursday and Tuesday
Sophie	I am free on Monday and Tuesday
Zoë	I am free on Friday
Robert	I am free on Tuesday and Wednesday
Stewart	I am free on Monday and Tuesday
Owen	I am free on Tuesday and Friday

Look at the days the children are free. Put your information in this table.

Day	Number of children who are free
Monday	
Tuesday	
Wednesday	
Thursday	
Friday	

Which day are most children free? _____

Which day are fewest children free? _____

How many more children can go on a Friday than a Wednesday?

_____

Which days would you be able to go on? _____

Organizing and using data

Lesson 4

Whole class activity

The children in class 1 have been looking at and playing different musical instruments. They are all going to choose which music lessons they would like to have.

This is a list of the children and the instruments they would like to learn to play.

Siân	Piano
Leonie	Recorder
Tao	Drum
Ross	Drum
Joe	Drum
Christopher	Piano

Can you organize the information in a different way?

Instrument	Number of children who like it
Piano	
Recorder	
Drum	

Can you build a block graph with your results?

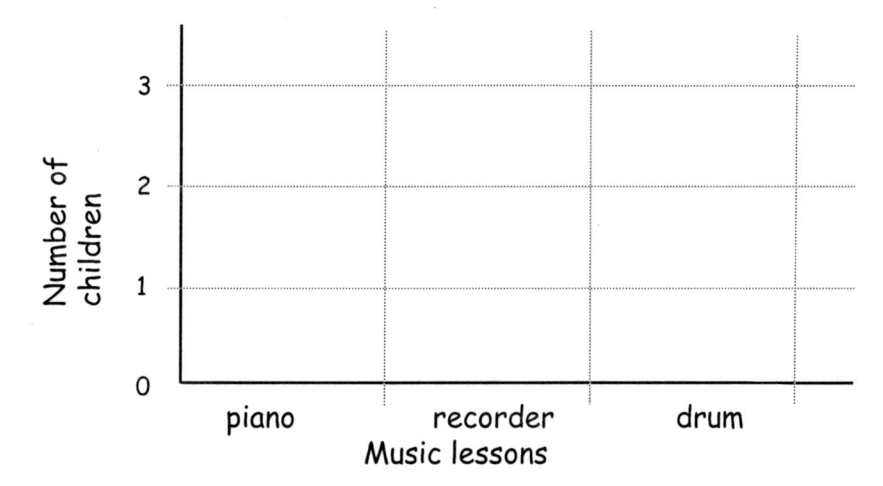

Which instrument is the most popular? _____
Which instrument is the least popular? _____
Will there be more children in the piano lessons or in the drum lessons? _____

Organizing and using data

Lesson 4a

Miss Young has decided to take her class on a school trip so she has asked them where they would like to go. This is a list of the children in her class and where they would like to go.

Rebecca	Seaside
Lloyd	Woodland walk
Indra	Museum
Abigail	Seaside
Faye	Woodland walk
Gina	Swimming pool
Natalie	Seaside

Can you organize the information in a different way?

Place	No. of children who want to go there
Seaside	
Woodland walk	
Swimming pool	
Museum	

Can you build a block graph with your results?

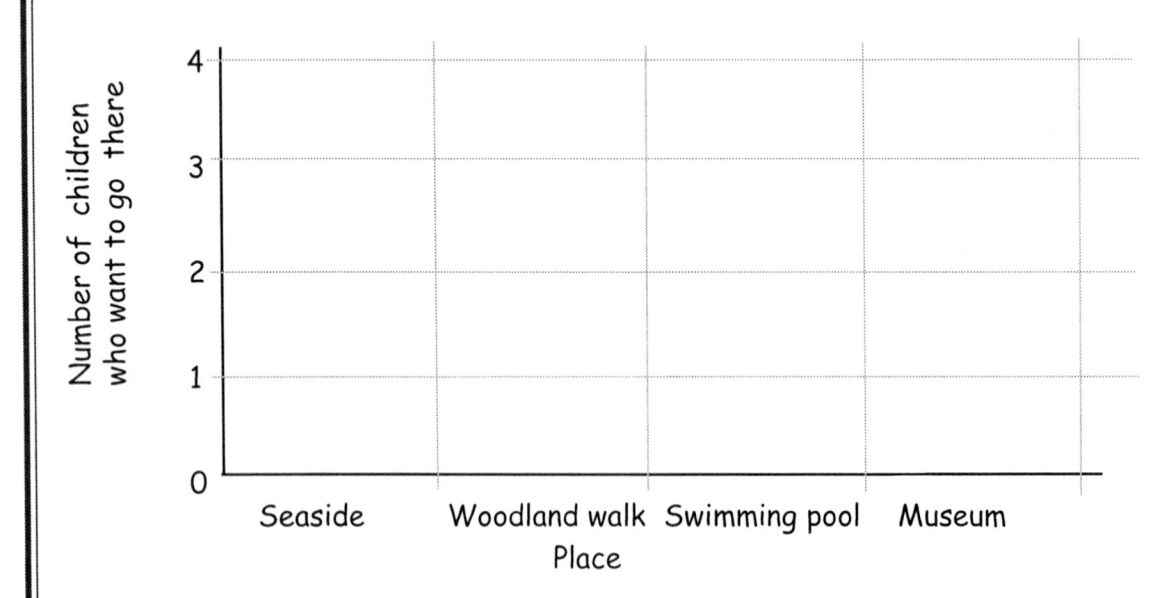

[Block graph with y-axis "Number of children who want to go there" (0-4) and x-axis "Place" with categories: Seaside, Woodland walk, Swimming pool, Museum]

Which place do most children want to go to? _____

Which place do fewest children want to go to? _____

Where would you like to go? _____

Organizing and using data

Lesson 4b

Miss Young has decided to take her class on a school trip so she has asked them where they would like to go. This is a list of the children in her class and where they would like to go.

Rebecca	Seaside	Natalie	Seaside
Lloyd	Woodland walk	Ryan	Woodland walk
Indra	Museum	Joanna	Museum
Abigail	Seaside	Marcus	Swimming pool
Faye	Woodland walk	Alexia	Swimming pool
Gina	Swimming pool		

Can you organize the information in a different way?

Place	Number of children who want to go there
Seaside	
Woodland walk	
Swimming pool	
Museum	

Can you build a block graph with your results?

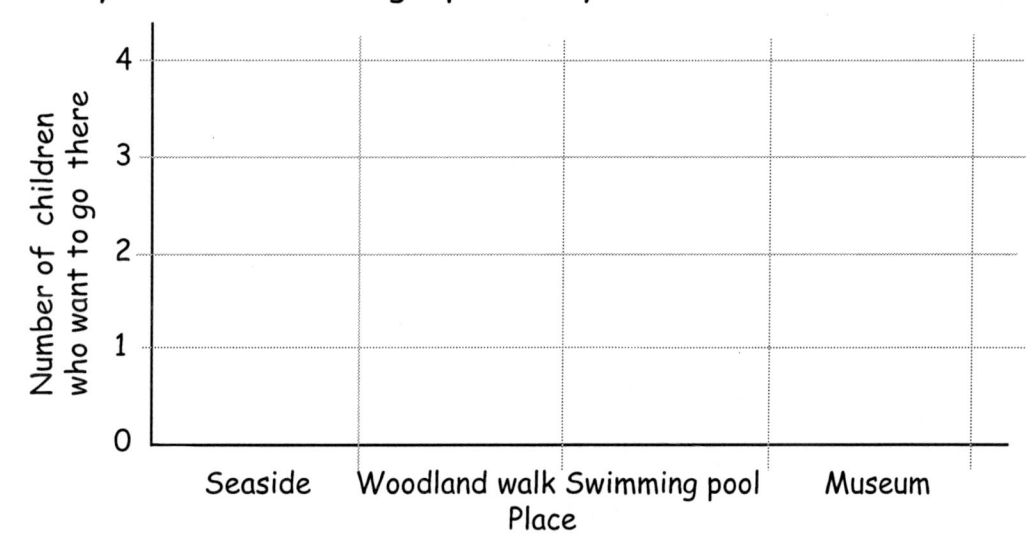

Which place do most children want to go to? _____
Which place do fewest children want to go to? _____
Where would you like to go? _____

Organizing and using data

Lesson 4C

Miss Young has decided to take her class on a school trip so she has asked them where they would like to go. This is a list of the children in her class and where they would like to go.

Rebecca	Seaside	Marcus	Museum
Lloyd	Woodland walk	Alexia	Seaside
Indra	Museum	Craig	Seaside
Abigail	Seaside	Tina	Woodland walk
Faye	Woodland walk	Mitchell	Woodland walk
Gina	Swimming pool	Molly	Swimming pool
Natalie	Seaside	Georgina	Swimming pool
Ryan	Woodland walk	Samuel	Seaside
Joanna	Museum		

Can you organize the information in a different way?

Place	No. of children who want to go there
Seaside	
Woodland walk	
Swimming pool	
Museum	

Can you build a block graph with your results?

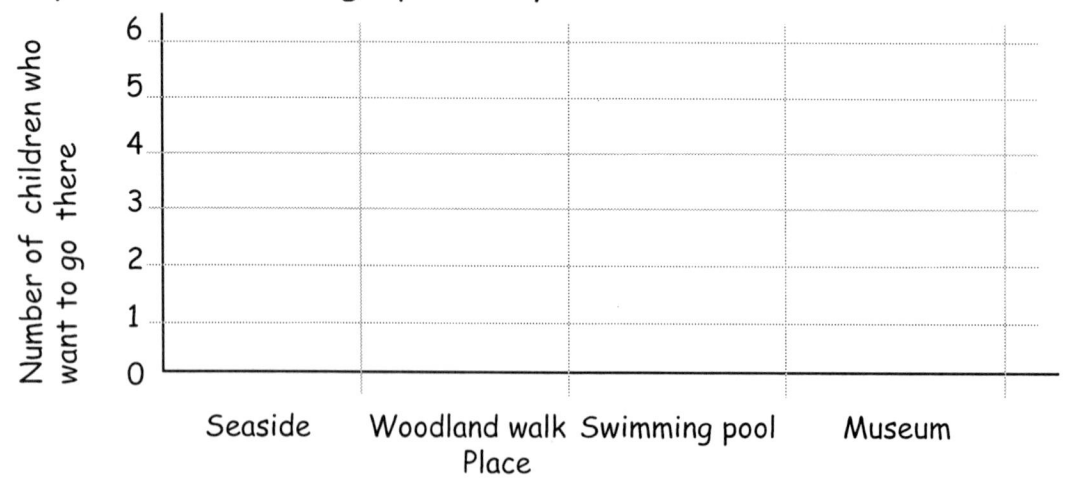

Which place do most children want to go to? _____
Which place do fewest children want to go to? _____
How many more children want to go the woodland walk than want to go swimming? _____
Where would you like to go? _____

Whole class activity

Organizing and using data

Lesson 5

Six children from the school took part as a team in a quiz. The team won. Now two of the children have to take part in the next round of the quiz but the head teacher doesn't know which children to send. The children in the team are called Yusef, Leo, Sharon, Frankie, Charlotte and Isabel.

Write down the names of the children in a list.

_____
_____
_____
_____
_____
_____

The children each answer 10 questions. Yusef gets 4 questions correct, Leo gets 8 questions correct, Sharon gets 4 questions correct, Frankie correctly answers one less than Leo. Charlotte gets right 2 more than Sharon. Isabel gets 5 questions correct.

Can you put this information into the table below?

Name	Number of correct answers

Who answers the most questions correctly? _____
Who gets the most questions wrong? _____
Which 2 children answer the same number of questions correctly? _____
Which 2 children should go on to the next round of the quiz? _____
Why? _____

Organizing and using data

Lesson 5a

The local shop has decided to sell ice-creams in the summer but they cannot decide which flavour they should sell. Make a list of three flavours which you think would be good to sell in the shop.

_____

_____

_____

Ask 6 children in your class which one of the ice-creams they would choose. Put their answers into this table. You could write the name of the flavours or you could use a colour.

Name	Ice-cream flavour

Count how many children want each sort of ice-cream. Put your answers in this table.

Flavour of ice-cream	Number of children who like it

Which ice-cream flavour should the shop sell?
_____

Why? _____

Organizing and using data

Lesson 5b

The local shop has decided to sell ice-creams in the summer but they cannot decide which flavour they should sell. Make a list of four flavours which you think would be good to sell in the shop.

_____
_____
_____
_____

Ask 8 children in your class which one of the ice-creams they would choose. Put their answers into this table. You could write the name of the flavours or you could use a colour.

Name	Ice-cream flavour

Count how many children want each sort of ice-cream. Put your answers in this table.

Flavour of ice-cream	Number of children who like it

Which ice-cream flavour should the shop sell? _____
Why? _____
Is there a flavour that you think would not sell very well?
_____
Why? _____

Organizing and using data

The local shop has decided to sell ice-creams in the summer but they cannot decide which flavour they should sell. Make a list of four flavours which you think would be good to sell in the shop.

_____

_____

_____

_____

Ask 10 children in your class which one of the ice-creams they would choose. Put their answers into this table. You could write the name of the flavours or you could use a colour.

Name	Flavour	Name	Flavour

Count how many children like each sort of ice-cream. Put your answers in this table.

Flavour of ice-cream	Number of children who like it

Which ice-cream flavour should the shop sell? _____

Why? _____

Is there a flavour that you think would not sell very well?

_____

Why? _____

Which ice-cream would you like? _____

Add your name to the list.

How many children like that ice-cream now? _____

Organizing and using data

Lesson 6

Whole class activity

Miss Martin, the school librarian, has decided that the school needs some new books but she cannot decide on the types of books to get. She has decided to let the children choose the books so she needs to find out what types of books they like.

This is a list of the children Miss Martin asks and the types of book they like to read.

Rachel	sport	Jonathan	art
Rupert	science	Sarah	art
Tanya	history	Amber	art
Robert	science	Victoria	science
Katie	art	Akshay	sport

Can you organize the information in a different way?

Type of books	Number of children who like it
sport	
science	
history	
art	

Can you build a block graph with your results?

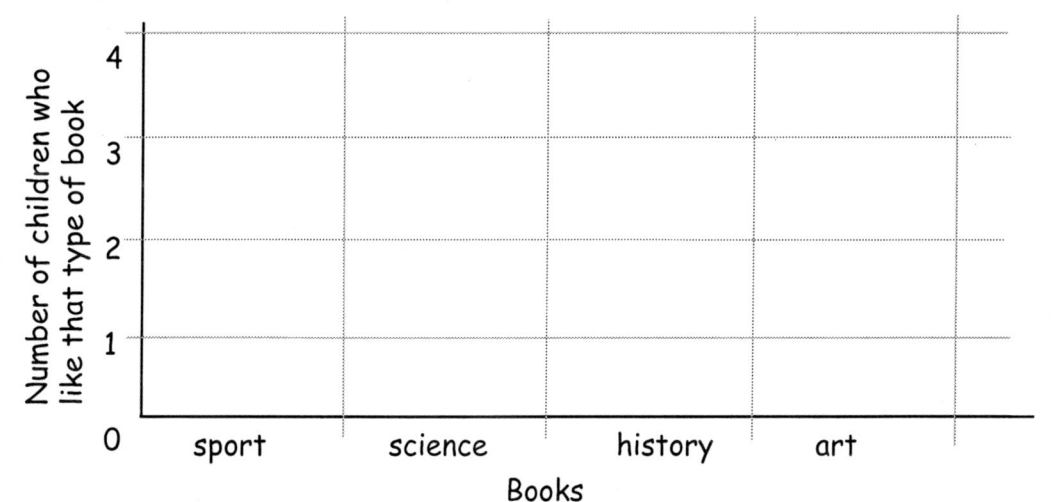

[Block graph: y-axis labelled "Number of children who like that type of book" from 0 to 4; x-axis labelled "Books" with categories: sport, science, history, art]

Do most children prefer the history books? _____

Do the children prefer the science books least? _____

Which type of books do you like? _____

Lesson 6a

Organizing and using data

Mr Evans, the head teacher, has decided that the school needs a new minibus but he cannot decide on the colour. He has decided to let the children choose the colour so he needs to find out what they like.

This is a list of the children in his class and their favourite colour for the minibus.

Bryony	Yellow
Paula	Orange
Caleb	Yellow
Ravi	Blue
Simone	Yellow
Belinda	Black

Can you organize the information in a different way?

Colour	Number of children who like it
Yellow	
Black	
Orange	
Blue	

Can you build a block graph with your results?

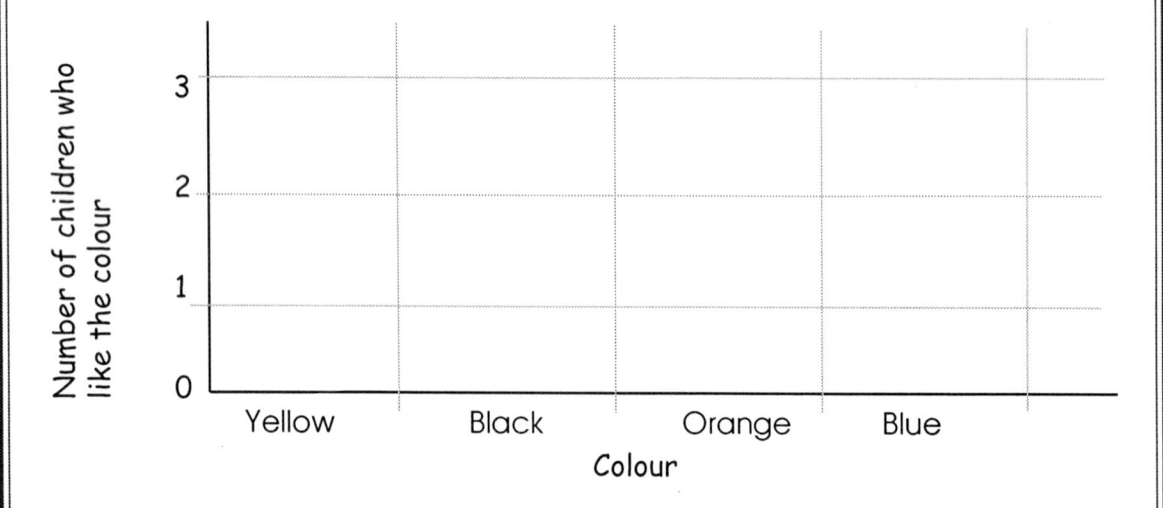

What colour minibus do you think Mr Evans should order?
_____

Which colour do you like? _____

Organizing and using data

Lesson 6b

Mr Evans, the head teacher, has decided that the school needs a new minibus but he cannot decide on the colour. He has decided to let the children choose the colour so he needs to find out what they like.

This is a list of the children in his class and their favourite colour for the minibus.

Bryony	Orange	Paula	Blue
Caleb	Orange	Ravi	Black
Simone	Yellow	Belinda	Yellow
Peter	Blue	Sanjeev	Blue
Bethan	Yellow	David	Blue

Can you organize the information in a different way?

Colour	Number of children who like it
Yellow	
Black	
Orange	
Blue	

Can you build a block graph with your results?

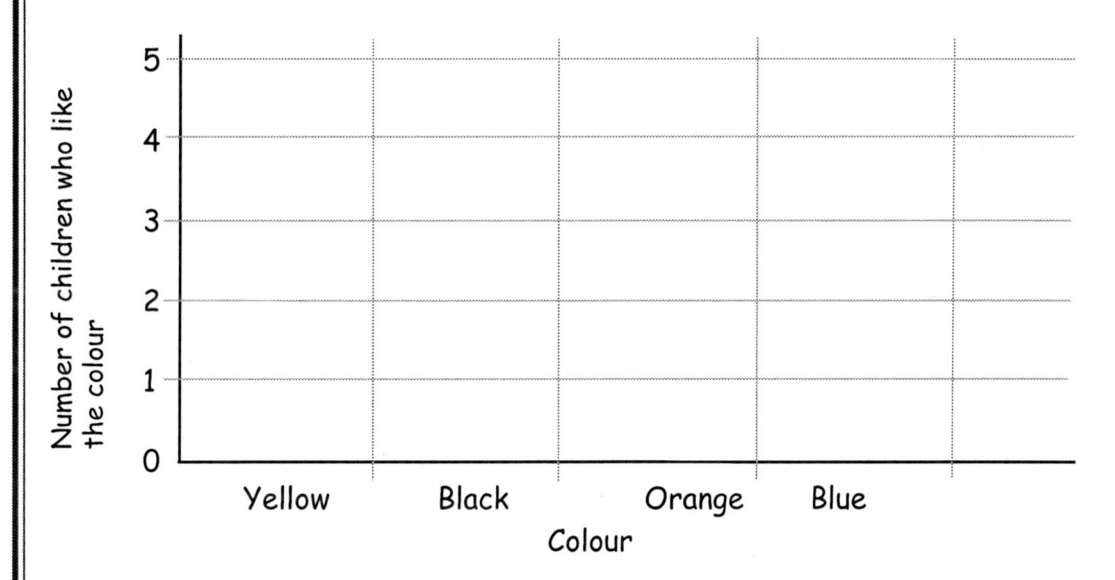

Which colour do most children like? _____
Which colour do they like least? _____
Which colour do you like? _____

Organizing and using data

Mr Evans, the head teacher, has decided that the school needs a new minibus but he cannot decide on the colour. He has decided to let the children choose the colour so he needs to find out what they like.

This is a list of the children in his class and their favourite colour for the minibus.

Bryony	Yellow	Bethan	Orange
Paula	Blue	David	Blue
Caleb	Orange	Imogen	Orange
Ravi	Blue	Lucy	Black
Simone	Yellow	Saul	Yellow
Belinda	Black	Reuben	Yellow
Peter	Orange	Isaac	Blue
Sanjeev	Blue	Susie	Blue

Can you organize the information in a different way?

Colour	Number of children who like it
Yellow	
Black	
Orange	
Blue	

Can you build a block graph with your results?

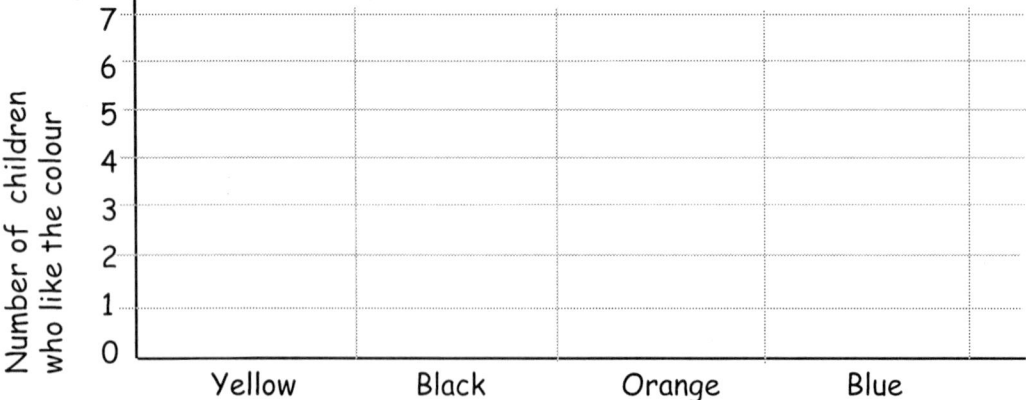

(Block graph: y-axis labelled "Number of children who like the colour" from 0 to 7; x-axis labels: Yellow, Black, Orange, Blue)

Which colour do most children like? _____

Which colour do they like least? _____

How many more children like yellow than blue? _____

Which colour do you like? _____

Answers

Making decisions

Lesson 1 (page 11)
A: 8 pencils; B: 16p; C: 2 hours
Lessons 1a–1c (pages 12–14)

Q	1a	1b	1c
1	3 grapes	4 grapes	7 grapes
2	£1	£3	£6
3	8 o'clock	7 o'clock	5 o'clock
4	4 cups	8 cups	12 cups

Lesson 2 (page 15)
A: 5 rulers; B: 4p; C: 6
Lessons 2a–2c (pages 16–18)

Q	2a	2b	2c
1	8 legs	12 legs	20 legs
2	10p	17p	24p
3	4 o'clock	5 o'clock	7 o'clock
4	4 jugs	4 jugs	4 jugs

Lesson 3 (page 19)
A: 3 hours; B: 7 pages; C: Any combination of coins that make 12p.
Lessons 3a–3c (pages 20–22)

Q	3a	3b	3c
1	3 metres	5 metres	6 metres
2	3	5	8
3	4 hours	6 hours	14 hours
4	8	15	23

Lesson 4 (page 23)
A: 16p; B: 9 marbles; C: 14 minutes
Lessons 4a–4c (pages 24–26)

Q	4a	4b	4c
1	6 metres	8 metres	12 metres
2	8	13	17
3	3 days	7 days	9 days
4	2	5	8

Lesson 5 (page 27)
A: 12; B: 13p; C: 6
Lessons 5a–5c (pages 28–30)

Q	5a	5b	5c
1	£3	£5	£8
2	7 rulers	14 rulers	21 rulers
3	5	8	17
4	8	12	20

Lesson 6 (page 31)
A: 15 minutes; B: 4 o'clock; C: 14 rows
Lessons 6a–6c (pages 32–34)

Q	6a	6b	6c
1	5 jugs	8 jugs	10 jugs
2	£8	£11	£18
3	9 cubes	17 cubes	21 cubes
4	10	14	22

Reasoning about numbers or shapes

Lesson 1 (page 35)
A: Class discussion, at least 8;
B: true; C: a square
Lessons 1a–1c (pages 36–38)

Q	1a	1b	1c
1	true	true	true
2	yes	yes	yes
3	5	8	20
4	● 8 ■ 4 ▲ 4	● 8 ■ 4 ▲ 2 ✚ 2	● 8 ■ 4 ▲ 2 ✚ 2 ♣ 6

107

Lesson 2 (page 39)

A: even; B: 9p; C: any pair of numbers with a difference of 3, for example 4 and 7 or 22 and 25.

Lessons 2a–2c (pages 40–42)

Q	2a	2b	2c
1	any combinations which total		
	8p	10p	15p
2	2–4 3–5 7–9	3–7 5–9 11–15	0–5 6–11 15–20
3	any number using the digits 1 and 2, eg. 1, 2, 11, 12, 21		
4	3, 6, 7, 9	4, 10, 12, 16	7, 16, 19, 25

Lesson 3 (page 43)

A: any pattern; B: yes; C: cube, cuboid

Lessons 3a–3c (pages 44–46)

Q	3a	3b	3c
1	true	true	true
2	yes	yes	yes
3	○	□	□
4	sphere, cylinder, cone	cube, cuboid	cube, cuboid, square-based pyramid, triangular prism

Lesson 4 (page 47)

A: yes; B: 2; C: 4, 6, 8, 10

Lessons 4a–4c (pages 48–50)

Q	4a	4b	4c
1	true	true	true
2	2	3	5
3	4	5	6
4	8 is more 2 is less 6 is more	9 is more 3 is less 6 is more	8 is less 17 is more 12 is more

Lesson 5 (page 51)

A: 4; B: any pattern; C: discussion

Lessons 5a–5c (pages 52–54)

Q	5a	5b	5c
1	true	true	true
2	true	true	true
3	○ 1 side △ 3 sides □ 4 sides	○ 1 side △ 3 sides □ 4 sides ▭ 4 sides	○ 1 side △ 3 sides □ 4 sides ▭ 4 sides ⬠ 5 sides
4	6	11	8

Lesson 6 (page 55)

A: 6 (8–2); B: any number; C: 3, 5, 7, 9, 11, 13

Lessons 6a–6c (pages 56–58)

Q	6a	6b	6c
1	true	true	true
2	true	true	true
3	5–9	4–11	5–13
4	40, 22, 31	15, 42, 33	71, 44, 53, 62

Problems involving 'real life', money or measures

Lesson 1 (page 59)
A: 9p; B: 16 hands; C: 5

Lessons 1a–1c (pages 60–62)

Q	1a	1b	1c
1	4	5	5
2	2p	4p	6p
3	4 metres	6 metres	9 metres
4	8p	13p	18p

Lesson 2 (page 63)
A: 6; B: 15; C: 6 o'clock

Lessons 2a–2c (pages 64–66)

Q	2a	2b	2c
1	8	9	11
2	any combination that adds up to		
	9p	12p	19p
3	3 mins.	7 mins.	10 mins.
4	6 pm	5 pm	4 pm

Lesson 3 (page 67)
A: 4 chairs; B: 12; C: 4p left, 2p + 2p or 2p + 1p + 1p or 1p + 1p + 1p +1p.

Lessons 3a–3c (pages 68–70)

Q	3a	3b	3c
1	7	12	13
2	14p	18p	22p
3	8	11	10
4	2 bricks	5 bricks	7 bricks

Lesson 4 (page 71)
A: 24; B: 7 minutes; C: 18 hands

Lessons 4a–4c (pages 72–74)

Q	4a	4b	4c
1	apple and orange	apple and banana	apple + orange
		banana and orange	orange + banana
2	4 spaces	6 spaces	9 spaces
3	10 feet	15 feet	17 feet
4	5p	8p	9p

Lesson 5 (page 75)
A: 12 sweets; B: 9; C: 9

Lessons 5a–5c (pages 76–78)

Q	5a	5b	5c
1	3	5	14
2	10 euros	15 euros	20 euros
3	6 mins.	11 mins.	14 mins.
4	6	8	11

Lesson 6 (page 79)
A: £4; B: 24p; C: 6 o'clock

Lessons 6a–6c (pages 80–82)

Q	6a	6b	6c
1	6	12	16
2	any combination which totals		
	10p	14p	20p
3	£5	£7	£10
4	4	8	11

Other Maths Problem Solving books in the series.

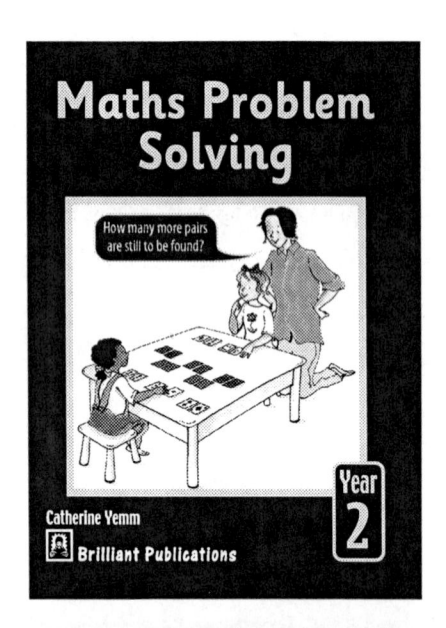

Other KS1 maths titles produced by Brilliant Publications.

A-Z Maths Games
Has over 50 games which reinforce maths skills and concepts. Create fun for pupils from the start with these highly interesting and self-motivating worksheets that focus on order of numbers, addition, subtraction, making graphs, measuring, telling time and greater than, less than and more than. These games are ideal for use during the numeracy hour.

Maths titles, How to Sparkle at... series
A wonderful collection of worksheets, activities and games for the numeracy hour. These books contain photocopiable, hands-on practical activities, work-alone activities and games that provide children with stimulating experiences, to help develop an understanding of the mathematical processes.

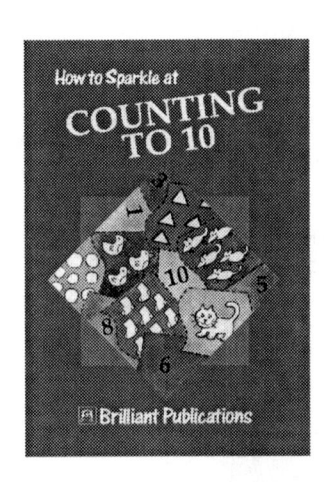

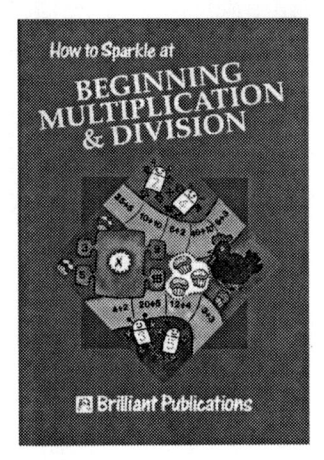

Printed in the United Kingdom
by Lightning Source UK Ltd.
130087UK00001B/101-150/A

9 781903 853740